RECOGNITION OF HUMANS AND THEIR ACTIVITIES USING VIDEO

Recognition of Humans and Their Activities Using Video
Rama Chellappa, Amit K. Roy-Chowdhury and Shaohua K. Zhou

ISBN: 978-3-031-01108-5 Chellappa, Recognition of Humans and Their Activities Using Video
(paperback)
ISBN: 978-3-031-02236-4 Chellappa, Recognition of Humans and Their Activities Using Video
(e-book)
DOI 10.1007/978-3-031-02236-4

Library of Congress Cataloging-in-Publication Data

First Edition
10 9 8 7 6 5 4 3 2 1

RECOGNITION OF HUMANS AND THEIR ACTIVITIES USING VIDEO

Rama Chellappa
Department of Electrical and Computer Engineering and
Center for Automation Research, UMIACS
University of Maryland,
College Park, MD, 20742, USA

Amit K. Roy-Chowdhury
Department of Electrical Engineering,
University of California,
Riverside, CA 92521, USA

S. Kevin Zhou
Integrated Data Systems Department
Siemens Corporate Research, Inc.
Princeton, NJ 08540, USA

ABSTRACT

The recognition of humans and their activities from video sequences is currently a very active area of research because of its applications in video surveillance, design of realistic entertainment systems, multimedia communications, and medical diagnosis. In this lecture, we discuss the use of face and gait signatures for human identification and recognition of human activities from video sequences. We survey existing work and describe some of the more well-known methods in these areas. We also describe our own research and outline future possibilities.

In the area of face recognition, we start with the traditional methods for image-based analysis and then describe some of the more recent developments related to the use of video sequences, 3D models, and techniques for representing variations of illumination. We note that the main challenge facing researchers in this area is the development of recognition strategies that are robust to changes due to pose, illumination, disguise, and aging. Gait recognition is a more recent area of research in video understanding, although it has been studied for a long time in psychophysics and kinesiology. The goal for video scientists working in this area is to automatically extract the parameters for representation of human gait. We describe some of the techniques that have been developed for this purpose, most of which are appearance based. We also highlight the challenges involved in dealing with changes in viewpoint and propose methods based on image synthesis, visual hull, and 3D models.

In the domain of human activity recognition, we present an extensive survey of various methods that have been developed in different disciplines like artificial intelligence, image processing, pattern recognition, and computer vision. We then outline our method for modeling complex activities using 2D and 3D deformable shape theory. The wide application of automatic human identification and activity recognition methods will require the fusion of different modalities like face and gait, dealing with the problems of pose and illumination variations, and accurate computation of 3D models. The last chapter of this lecture deals with these areas of future research.

KEYWORDS

Pattern Recognition, Face Recognition, Gait Recognition, Human Activity Recognition

Contents

Acknowledgment

The material presented in this lecture is the result of research spanning a number of years and involving many researchers besides the authors. We would like to acknowledge support from DARPA's Human Identification (HID) and Human Activity Inference (HAI) projects under Grants N00014-00-1-0908 and N00014-02-1-0809, respectively, and NSF-ITR Grant-0325119. Amit Roy-Chowdhury would also like to acknowledge support from the University of California, Riverside. Volker Kruegar (Aalborg Universty, Denmark) was involved in some of the work in face and gait recognition. Amit Kale (University of Kentucky) and A. Rajagopalan (IIT Madras, India) developed the initial HMM-based gait recognition framework. Aravind Sundaresan (University of Maryland) worked with Chellappa and Roy-Chowdhury to modify the HMM framework which led to improved performance, as well as to design the kinematic chain model for human motion. Amit Kale was also involved in developing the view invariant gait recognition algorithm. Naresh Cuntoor and N. Ramanathan (both University of Maryland) participated in testing the gait recognition algorithms on standard datasets. Namrata Vaswani (Georgia Tech) and Ashok Veeraraghavan (University of Maryland) worked on the shape-dynamical models for activity and gait representation. Umut Akdemir (Siemens) was involved in the work on 3D models for gait recognition. The section on review of face recognition in Chapter 2 has been excerpted from a recent survey paper (194).

CHAPTER 1

Introduction

Computer vision broadly refers to the discipline where extraction of useful 2D and/or 3D information from one or more images is of interest. Since the human visual system works by extracting information from the images formed on the retina of the eye, developments in computer vision are inevitably compared to the abilities of the human vision system. One of the basic tasks of the human visual system is to recognize humans and objects and spatial relationships among them. Similarly, one of the main goals of computer vision researchers is to develop methods for localization and recognition of objects in a scene. A special case of this general problem is the recognition of humans and their activities. In this monograph, we outline some of the methods that have been recently developed toward achieving this goal, and describe our present research in this area. Specifically, we divide the recognition problem in two parts: human identification using face and gait, and human activity recognition.

The basic input for recognition systems is a video sequence (a single image being a special case of it). Unless otherwise specified, we will assume throughout that the input data lies in the visible spectrum. However, video is collected under different conditions and it is unrealistic to expect the same kind of performance even as the resolution of the images varies. Face recognition algorithms usually require high-resolution video where the face is the predominant subject in each frame. Gait recognition can work with lower resolution data, provided one can extract the motion of the different parts of the human body. Our experience is that video data on which gait recognition algorithms perform reasonably well, usually do not have the resolution for satisfactory performance with face recognition algorithms. In many practical situations, the activity being performed is often of more importance that the identity of the person involved. Also, the input data may be of such a resolution that it is impossible to identify any single individual, but it may be possible to understand the tasks being carried out by a group of people. Under such circumstances, activity recognition algorithms are required.

Traditionally, problems in computer vision have been grouped into three areas that have vaguely defined boundaries. At the so-called low level, the goal is to extract features such as edges, corners, lines, segmented regions, and track features over a sequence of frames or to compute optical flow. At the intermediate level, using the output of the low-level modules, one is interested in grouping of features, in estimation of depth using stereopsis, and in motion and structure estimation. At the high level, the intermediate-level outputs are combined with available knowledge about the scene, objects, and tasks so that descriptions of objects can be derived.

Recognition algorithms usually require a combination of various techniques, which span across all the three levels. At the lowest level are methods for localizing the area of interest in the image (background subtraction), detection and tracking of feature points, various morphological operations that may be needed in order to obtain a better quality of the input data, etc. Middle-level operations usually involve methods for 3D reconstruction, view synthesis, system modeling, and the estimation of parameters from the input data. High-level techniques are used for the actual recognition task based on the information of the various parameters learned at the mid-level. Thus, a recognition problem can be described as a geometric inference problem since it aims to obtain an understanding of the 3D world that we live in from 2D images of it. As an example, consider that we are given a video sequence of a person performing different activities, like walking, sitting, running, etc. Our goal is to identify the particular activity. Given the video sequence, we would first like to extract low-level features such as motion parameters, edge descriptors, segmentation regions, etc. Using this data, we may want to build some kind of a model that is specific to the activity, e.g. a 3D representation of it so that the activity can be recognized from any viewing direction. Finally, we would like to compare the model parameters learned for the different activities and compare them in order to obtain the correct recognition. At this stage, we may also need to string together various activities, using a proper logical framework, in order to deal with a sequence of activities.

1.1   OVERVIEW OF RECOGNITION ALGORITHMS

We will now provide a brief historical overview of the three areas that we deal with in this book. The aim is to explain the challenges underlying the different tasks, as well as the successes that have been achieved. A detailed review of various techniques will be provided in each of the chapters.

Between face, gait, and activity, probably the maximum amount of effort has been devoted to the problem of face recognition. The problem can be defined as follows. A database of a large number of faces is available as the gallery. The faces may be represented as a single image or a set of images, either as a video sequence or a collection of discrete poses. These images are usually referred to as training images, since they are used to train the parameters of a recognition algorithm. Given an image or a set of images of an individual (known as test images), the problem is to identify the individual from the gallery or decide that he/she is not part of the gallery. The main challenges facing computer vision researchers in this problem are

- varying conditions of illumination between the training and test images;

- changes in appearance, make-up, and clothing between the training and test images;

- changes due to difference in time between the recording of the training and test images, leading to aging effects;

- different poses of the face in different instances of recording.

All of these issue make face recognition an extremely challenging problem. However, considerable progress has been made in the last decade and face recognition technologies are under consideration for deployment at various facilities.

Chronologically speaking, face recognition first started with still images. Popular methods that have been proposed are principal components analysis or eigenfaces (113; 173), linear discriminant analysis or Fisherfaces (10; 193), elastic graph matching (184), local feature analysis (126), morphable models (15), and numerous others discussed in detail in Chapter 2. While recognition rates under controlled indoor situations are reasonably good, a lot needs to be done before such technologies can be deployed in outdoor situations. Many researchers believe that the use of video sequences, as opposed to a single image, will lead to much better recognition rates. This is based on the intuition that integrating the recognition performance over a sequence would give a better result than considering just one single image from that sequence. Therefore, most of the present research in this area is in this direction.

Gait recognition is helpful because lower resolution images can be used, making the method less intrusive and may not require the active cooperation of the subject. The solution of problem of human motion modeling has very important implications for different areas like surveillance, medical diagnosis, entertainment industry, video communications, etc. Traditionally, there has been a significant interest in studying human motion in various disciplines. In psychology, Johansson conducted classic experiments by attaching light displays to various body parts and showed that humans can identify motion when presented with only a small set of these moving dots (76). Muybridge captured the first photographic recordings of humans and animals in motion in his famous publication on animal locomotion toward the end of the nineteenth century (116). In kinesology the goal has been to develop models of the human body that explain how it functions mechanically (64). Gait recognition is a relatively new area to computer vision researchers. However, some progress has been made and reasonably good performance on large datasets under controlled circumstances has been achieved. The problems facing this area are often similar to the ones in face recognition—poor performance in uncontrolled outdoor situations and the effects of time. Some progress has also been made in recognizing people walking in arbitrary directions to the camera.

The third area that this book deals with is the recognition of human activities. The challenges to developing robust activity recognition algorithms are the large number of activities, representation issues, and metrics for determining how close an activity is to the model activities. In order to recognize different activities, it is necessary to construct an ontology of various normal (both frequent and rare) events. Deviations from a pre-constructed dictionary can then be classified as abnormal events. It is also necessary that the representation be invariant to the viewing direction of the camera, and independent of the number of cameras (i.e. should be scalable to a video sensor network). Trajectories, usually computed from 2D video data, are a natural starting point for activity recognition systems. Trajectories contain a lot of information about the underlying event that they represent. However, most prevalent systems do little more than tracking a set of points over a sequence of images, and try to infer about the event from the set of tracks. Trajectories are ambiguous (different events can have the same trajectory) and depend on the viewing direction. Also, identifying events from trajectories requires the enunciation of a set of rules (often ad-hoc), which can vary from one instance to another of the same event. However, there are methods that have been developed recently on modeling activities based on information available in the trajectories. We will describe two such methods that we have developed recently, which use the 2D and 3D shape in order to recognize an activity.

As we understand these processes better, the limitations of some of the existing algorithms become more apparent. Also, progress in allied fields like image/video processing, pattern recognition, phychophysics, and neuroscience allows us to develop more realistic (and often more complicated) mathematical models for the underlying physical processes. We will explore some of the advanced techniques toward the end of this monograph. Specifically, we will discuss the use of 3D face models estimated from motion information and kinematic chain models for human motion modeling in a multicamera framework, where the kinematic parameters are estimated from video sequences.

In the next chapter, we deal with methods for face recognition. We start with the classical image-based recognition techniques, like eigenfaces, and move to the more recent approaches using morphable models, photometric stereo, and video sequences. An experimental comparison of the various methods is also presented. Chapter 3 deals with gait recognition methods. Three techniques are described in detail: Hidden Markov Model (HMM) based recognition, view-invariant recognition from monocular video, and shape-dynamical models for gait representation. Our description of activity recognition in Chapter 4 begins with a survey of the various methods that have been developed over the last few deacdes in this area. We then move on to describe our recent work on computational models using the dynamics of 2D and 3D shape sequences for representation of activities. The monograph concludes by discussing some future research directions in these areas, along with existing results that will form the basis for the research.

CHAPTER 2

Human Recognition Using Face

As one of the most successful applications of image analysis and understanding, face recognition has recently received significant attention, especially during the past decade. This is evidenced by the emergence of face recognition conferences such as AVBPA (2) and AFGR (1), systematic empirical evaluations of face recognition techniques (FRT), including the FERET (131), FRVT 2000 (14), FRVT 2002 (130), and XM2VTS (111) protocols, and many commercially available systems. There are at least two reasons for this trend; the first is the wide range of commercial and law enforcement applications and the second is the availability of feasible technologies after 30 years of research. Applications of face recognition technology (FRT) range from static, controlled-format photographs to uncontrolled video images, posing a wide range of technical challenges and requiring an equally wide range of techniques from image processing, analysis, understanding, and pattern recognition.

One can broadly classify FRT systems into two groups depending on whether they make use of static images or video. A general statement of the problem of machine recognition of faces can be formulated as follows: given still or video images of a scene, identify or verify one or more persons in the scene using a stored database of faces. Available collateral information such as race, age, gender, facial expression, or speech may be used in narrowing the search (enhancing recognition). The solution to the problem involves segmentation of faces (face detection) from clut-tered scenes, feature extraction from the face regions, and identification or verifica-tion. In identification problems, the input to the system is an unknown face, and the system reports back the determined identity from a database of known individuals, whereas in verification problems, the system needs to confirm or reject the claimed identity of the input face. Because of the wide variety of problems involved, ma-chine recognition of human faces continues to attract researchers from disciplines such as image processing, pattern recognition, neural networks, computer vision, computer graphics, and psychology.

2.1 LITERATURE REVIEW

Over the past 30 years extensive research has been conducted by psychophysicists, neuroscientists and engineers on various aspects of face recognition by humans and machines. Psychophysicists and neuroscientists have been concerned with issues such as whether face perception is a dedicated process (11; 43; 52) and whether it is done holistically or by local feature analysis. The earliest results on automatic ma-chine recognition of faces can be traced back to the seminal work of Kanade (82) and Kelly (83) in the 1970s. Early approaches treated face recognition as a 2D pattern recognition problem, using measured attributes of features (e.g. the dis-tances between important points) in faces or face profiles (16; 82; 83). During the 1980s, work on face recognition in the computer vision community remained largely dormant. However, since the early 1990s, research interest in FRT has grown significantly.

Face recognition can be classified as holistic approaches, which consider the whole face at a time, or feature-based methods, which look at the interplay between the different features on the face. Among appearance-based holistic approaches, eigenfaces (85; 173) and Fisherfaces (10; 44; 193) have proved to be effective in experiments with large databases. Feature-based graph matching approaches (184) have also been quite successful. Compared to holistic approaches, feature-based methods (33) are less sensitive to variations in illumination and viewpoint and to inaccuracy in face localization. However, the feature extraction techniques needed for this type of approach are still not reliable or accurate enough. For example, most eye localization techniques assume some geometric and textural models and do not work if the eye is closed.

Recently, much research has concentrated on video-based face recognition (104; 200). The still image problem has several inherent advantages and disadvantages. For applications such as drivers' licenses, due to the controlled nature of the image acquisition process, the segmentation problem is rather easy. However, if only a static picture of an airport scene is available, automatic location and segmentation of a face could pose serious challenges to any segmentation algorithm. On the other hand, if a video sequence is available, segmentation of a moving person can be more easily accomplished using motion as a cue. But the small size and low image quality of faces captured from video can significantly increase the difficulty in recognition.

A thorough review of the literature in face recognition is available in (194). During the past 8 years, face recognition has received increased attention and has advanced technically. Many commercial systems for still face recognition are now available. Recently, significant research efforts have been focused on video-based face modeling/tracking, recognition, and system integration. New data sets have been created and evaluations of recognition techniques using these databases have been carried out. It is not an overstatement to say that face recognition has become one of the most active applications of pattern recognition, image analysis, and understanding.

2.2   FACE RECOGNITION FROM STILL IMAGES

The problem of automatic face recognition involves three key steps/subtasks: (1) Detection and coarse normalization of faces, (2) feature extraction and accurate normalization of faces, and (3) identification and/or verification. Sometimes, different subtasks are not completely independent. Face detection and feature extraction can often be achieved simultaneously.

2.2.1   Preprocessing Steps—Face Detection and Feature Extraction

Prior to feeding an image to an automatic face recognition algorithm, it is necessary to detect the face in each image and extract relevant features. Most of the early work on face detection involved single-face segmentation from a simple or complex background. These approaches included using a whole-face template, a deformable feature-based template, skin color, and a neural network. Significant advances have been made in recent years in achieving automatic face detection under various conditions. Compared to feature-based methods and template-matching methods, appearance- or image-based methods (140; 164) that train machine systems on large numbers of samples have achieved the best results. More recently, detection of faces under rotation in depth has been studied. One approach is based on training on multiple-view samples (62; 146). Compared to invariant-feature-based methods (184), multiview-based methods of face detection and recognition seem to be able to achieve better results when the angle of out-of-plane rotation is large (35°).

Three types of feature extraction methods can be distinguished: (1) Generic methods based on edges, lines, and curves; (2) feature-template-based methods that are used to detect facial features such as eyes; (3) structural matching methods that take into consideration geometrical constraints on the features. Early approaches focused on individual features. These methods have difficulty when the appearances of the features change significantly, e.g., closed eyes, eyes with glasses, open mouth. To detect the features more reliably, recent approaches use structural matching methods, e.g., the Active Shape Model (32). Compared to earlier methods, these recent statistical methods are much more robust in terms of handling variations in image intensity and feature shape.

A method for facial feature detection using 2D shapes was presented in (114). The authors defined an operator for shape detection by applying the derivative of the double exponential function along the shape's boundary contour. This method of detecting shape was an extension of the task of edge detection at the pixel level to the problem of global contour detection.

2.2.2   Face Recognition—Broad Classification

The different techniques for face recognition can be grouped into the following broad categories.

1. Holistic matching methods: These methods use the whole face region as the raw input to a recognition system. One of the most widely used representations of the face region is eigenpictures (85; 173), which are based on principal component analysis [i.e. Karhunen–Loeve (KL) transform].

2. Feature-based (structural) matching methods: Typically, in these methods, local features such as the eyes, nose, and mouth are first extracted and their locations and local statistics (geometric and/or appearance) are fed into a structural classifier.

3. Hybrid methods: These methods combine both local features and the whole face region to recognize a face. One can argue that they could potentially offer the best of the above two types of methods.

We will now describe some of the more well-known approaches for face recognition from images.

2.2.3   Principal Component Analysis/Eigenfaces

Starting from the successful low-dimensional reconstruction of faces using KL or principal component analysis (PCA) projections (85; 173), eigenpictures have been one of the major driving forces behind face representation, detection, and recognition. It is well known that there exist significant statistical redundancies in natural images. For a limited class of objects such as face images that are normalized with respect to scale, translation, and rotation, the redundancy is even greater. One of the best global compact representations is KL/PCA, which decorrelates the outputs.

More specifically, sample vectors $\mathbf{x}$ can be expressed as linear combinations of the orthogonal basis Φ_i: $\mathbf{x} = \sum_{i=1}^{n} a_i \Phi_i \approx \sum_{i=1}^{m} a_i \Phi_i$ (typically $m \ll n$) by solving the eigenproblem

$$C\Phi = \Phi\Lambda, \tag{2.1}$$

where C is the covariance matrix for input $\mathbf{x}$.

Each face image is represented as a vector, $\mathbf{x}$. Given a gallery of training images, an eigenspace of faces is created. For each training image, the projections onto the eigenspace are learned. The projection vector for each image can be regarded as the biometric code for that particular face. Given a test image (where the face has been properly localized), the projections are obtained and compared with the projections of each of the faces in the gallery. The closest match (using a proper metric) identifies the person in the test image (see Figure 2.2). By using the observation that the projection of a face image and a non-face image are usually different, a method of detecting the presence of a face in a given image is obtained. The first really successful demonstration of machine recognition of faces was made in (173) using eigenpictures (also known as eigenfaces) for face detection and identification. The method was demonstrated using a database of 2500 face images of 16 subjects, in all combinations of three head orientations, three head sizes, and three lighting conditions.

Using a probabilistic measure of similarity, instead of the simple Euclidean distance used with eigenfaces (173), the standard eigenface approach was extended (113) to a Bayesian approach. Practically, the major drawback of a Bayesian method is the need to estimate probability distributions in a high-dimensional space from very limited numbers of training samples per class. To avoid this problem, a much simpler two-class problem was created from the multiclass problem by using a similarity measure based on a Bayesian analysis of image differences. Two mutually exclusive classes were defined: Ω_I, representing *intrapersonal* variations between multiple images of the same individual and Ω_E, representing *extrapersonal* variations due to differences in identity. Assuming that both classes are Gaussian

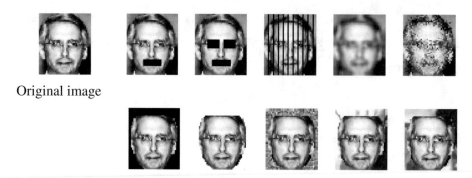

Original image

FIGURE 2.1: Electronically modified images that were correctly identified.

distributed, likelihood functions $P(\Delta \mid \Omega_I)$ and $P(\Delta \mid \Omega_E)$ were estimated for a given intensity difference $\Delta = I_1 - I_2$. Given these likelihood functions and using the MAP rule, two face images are determined to belong to the same individual if $P(\Delta \mid \Omega_I) > P(\Delta \mid \Omega_E)$. A large performance improvement of this probabilistic matching technique over standard nearest-neighbor eigenspace matching was reported using large face data sets including the FERET database (131).

2.2.4 Linear Discriminant Analysis/Fisher Faces

Face recognition systems using linear/Fisher discriminant analysis have also been very successful (10; 44; 166; 193). Linear discriminant analysis (LDA) training is carried out via scatter matrix analysis (50). For an M-class problem, the within- and

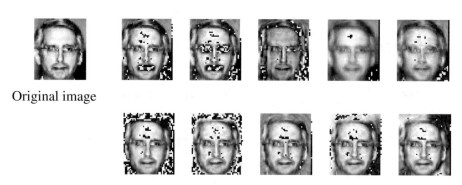

Original image

FIGURE 2.2: Reconstructed images using 300 PCA projection coefficients for electronically modified images (see also Fig. 2.1) (191).

between-class scatter matrices S_w, S_b are computed as follows:

$$S_w = \sum_{i=1}^{M} \Pr(\omega_i) S_{w,i} \tag{2.2}$$

$$S_b = \sum_{i=1}^{M} \Pr(\omega_i) S_{b,i},$$

where $\Pr(\omega_i)$ is the prior class probability, and is usually replaced by $1/M$ in practice with the assumption of equal priors. Here $S_{w,i} = E[(\mathbf{x}(\omega) - \mathbf{m}_i)(\mathbf{x}(\omega) - \mathbf{m}_i)^T \mid \omega = \omega_i]$ is the *Within-class Scatter Matrix*, showing the average scatter of the sample vectors $\mathbf{x}$ of the different classes ω_i around their respective means $\mathbf{m}_i$, and $S_{b,i} = (\mathbf{m}_i - \mathbf{m}_0)(\mathbf{m}_i - \mathbf{m}_0)^T$ is the *Between-class Scatter Matrix*, representing the scatter of the conditional mean vectors $\mathbf{m}_i$ around the overall mean vector $\mathbf{m}_0$. A commonly used measure for quantifying discriminatory power is the ratio of the determinant of the between-class scatter matrix of the projected samples to the determinant of the within-class scatter matrix: $\mathcal{J}(T) = \mid T^T S_b T \mid / \mid T^T S_w T \mid$. The optimal projection matrix W, which maximizes $\mathcal{J}(T)$, can be obtained by solving a generalized eigenvalue problem:

$$S_b W = S_w W \Lambda_W. \tag{2.3}$$

In (165; 166), discriminant analysis of eigenfeatures was applied in an image retrieval system to determine not only class (human face vs. non-face objects) but also individuals within the face class. A set of 800 images was used for training; the training set came from 42 classes, of which human faces belong to a single class. Within the single face class, 356 individuals were included and distinguished. Testing results on images not in the training set were 91% for 78 face images and 87% for 38 non-face images based on the top choice.

A comparative performance analysis was carried out in (10). Four methods were compared in this paper: (1) a correlation-based method, (2) a variant of the linear subspace method suggested in (150), (3) an eigenface method (173), and

(4) a Fisherface method, which uses subspace projection prior to LDA projection to avoid the possible singularity in S_w as in (166). The results of the experiments showed that the Fisherface method performed significantly better than the other three methods. However, no claim was made about the relative performance of these algorithms on larger databases.

LDA is a single-exemplar method in the sense that each class during classification is represented by a single exemplar, i.e. the sample mean of the class. The underlying assumption of LDA is that each class possesses a normal density with a different mean vector but a common covariance matrix. Under the above assumption, LDA coincides with the optimal Bayes classifier. However, face appearance lies in a highly complex manifold. To represent such a manifold, a large amount of exemplars are required. When in reality there is only a small number of samples per class to represent a complex manifold, all samples should be used as exemplars for a final classification. To this end, a multiple-exemplar discriminant analysis (MEDA) was proposed in (197).

The essence of MEDA is to redefine the within- and between-class scatter matrices S_w and S_b in (2.3)

$$S_w = \sum_{i=1}^{M} \Pr(\omega_i)\Pr(\omega_i)$$

$$S_{w,ii}; \quad S_{w,ii} = E[(\mathbf{x}(\omega) - \mathbf{x}(\omega'))(\mathbf{x}(\omega) - \mathbf{x}(\omega'))^T \,|\, \omega = \omega_i, \omega' = \omega_i], \quad (2.4)$$

$$S_b = \sum_{i,j=1; i \neq j}^{M} \Pr(\omega_i)\Pr(\omega_j)$$

$$S_{b,ij}; \quad S_{b,ij} = E[(\mathbf{x}(\omega) - \mathbf{x}(\omega'))(\mathbf{x}(\omega) - \mathbf{x}(\omega'))^T \,|\, \omega = \omega_i, \omega' = \omega_j]. \quad (2.5)$$

The same generalized eigenvalue problem as in Eq. (2.3) is solved to find the projection matrix W. The classification is to find the minimum distance to all exemplars (instead of the class mean) in the projected subspace. In (197), five subspace methods possessing a discriminative capability are compared on a subset of FERET database with 200 subjects and 3 variations (neutral face, smile expression, and illumination variation). The five approaches are LDA [10], subspace LDA [171], Bayes face recognition (112), intra-personal space (112), and MEDA. The MEDA approach yielded the highest recognition performance since it performs a discriminant analysis with multiple-exemplar modeling embedded.

2.2.5   Other Holistic Approaches

To improve the performance of LDA-based systems, a regularized subspace LDA system that unifies PCA and LDA was proposed in (191; 193). Good generalization ability of this system was demonstrated by experiments that carried out testing on new classes/individuals without retraining the PCA bases Φ, and sometimes the LDA bases W.

An evolution pursuit (EP) based adaptive representation and its application to face recognition was presented in (103). In analogy to projection pursuit methods, EP seeks to learn an optimal basis for the dual purpose of data compression and pattern classification. In order to increase the generalization ability of EP, a balance is sought between minimizing the empirical risk encountered during training and narrowing the confidence interval for reducing the guaranteed risk during future testing on unseen data (174). Toward that end, EP implements strategies characteristic of genetic algorithms (GAs) for searching the space of possible solutions to determine the optimal basis. EP starts by projecting the original data into a lower-dimensional whitened PCA space. Directed random rotations of the basis vectors in this space are then searched by GAs, where evolution is driven by a fitness function defined in terms of performance accuracy (empirical risk) and class separation (confidence interval). The feasibility of this method has been demonstrated for face recognition, where the large number of possible bases requires a greedy search algorithm. The particular face recognition task involves 1107 FERET frontal face images of 369 subjects; there were three frontal images for each subject, two for training and the remaining one for testing. The authors reported improved face recognition performance as compared to eigenfaces (173), and better generalization capability than Fisherfaces (10).

A fully automatic face detection/recognition system based on a neural network was reported in (101). The proposed system was based on a probabilistic decision-based neural network (PDBNN), which consisted of three modules: a face detector, an eye localizer, and a face recognizer.

2.2.6   Feature-Based Approaches: Elastic Bunch Graph Matching

One of the most successful methods among feature-based structural matching approaches is the elastic bunch graph matching (EBGM) system (184), which is based on dynamic link architecture (DLA) (91). Wavelets, especially Gabor wavelets, play a building block role for facial representation in these graph matching methods. A typical local feature representation consists of wavelet coefficients for different scales and rotations based on fixed wavelet bases (called "jets"). These locally estimated wavelet coefficients are robust to illumination change, translation, distortion, rotation, and scaling.

The basic 2-D Gabor function and its Fourier transform are

$$g(x, y : u_0, v_0) = \exp(-[x^2/2\sigma_x^2 + y^2/2\sigma_y^2] + 2\pi i[u_0 x + v_0 y]) \qquad (2.6)$$
$$G(u, v) = \exp(-2\pi^2(\sigma_x^2(u - u_0)^2 + \sigma_y^2(v - v_0)^2)),$$

where σ_x and σ_y represent the spatial widths of the Gaussian and (u_0, v_0) is the frequency of the complex sinusoid.

The DLA's basic mechanism consists of the connection parameter T_{ij} between two neurons (i, j) and a dynamic variable J_{ij}. The T-parameters can be changed slowly by long-term synaptic plasticity. The weights J_{ij} are subject to rapid modification and are controlled by the signal correlations between neurons i and j. Negative signal correlations lead to a decrease and positive signal correlations lead to an increase in J_{ij}. In the absence of any correlation, J_{ij} slowly returns to a resting state, a fixed fraction of T_{ij}. Each stored image is formed by picking a rectangular grid of points as graph nodes. The grid is properly positioned over the image and is stored with each grid point's locally determined jet [Fig. 2.3(a)], and serves to represent the pattern classes. Recognition of a new image takes place by

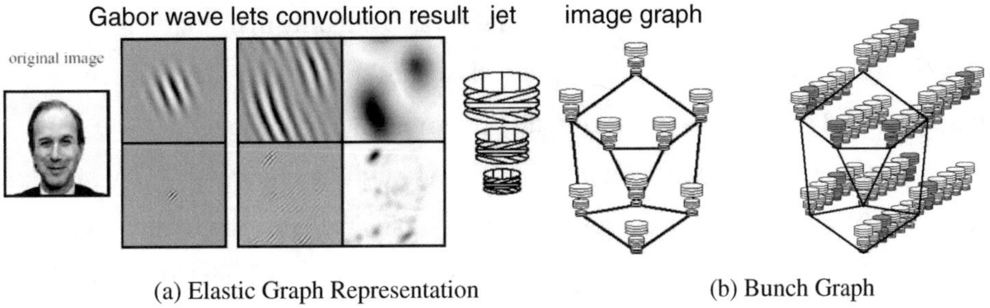

(a) Elastic Graph Representation (b) Bunch Graph

FIGURE 2.3: The bunch graph representation of faces used in elastic graph matching (184). (Courtesy of L. Wiskott, J.-M. Fellous, and C. von der Malsburg.)

transforming the image into the grid of jets, and matching all stored model graphs to the image. Confirmation of the DLA is done by establishing and dynamically modifying links between vertices in the model domain.

The DLA architecture was later extended to EBGM (184) (Fig. 2.3). This is similar to the graph described above, but instead of attaching only a single jet to each node, the authors attach a set of jets (called the bunch graph representation, Fig. 2.3(b)), each derived from a different face image. To handle the pose variation problem, the pose of the face is first determined using prior class information, and the "jet" transformations under pose variation are learned (89; 108). Systems based on the EBGM approach have been applied to face detection, feature finding, pose estimation, gender classification, sketch-image-based recognition, and general object recognition. The success of the EBGM system may be due to its resemblance to the human visual system (11).

2.2.7 Hybrid Approaches

Hybrid approaches use both holistic and local features. For example, the concept of eigenfaces can be extended to eigenfeatures, such as eigeneyes, eigenmouth, etc. Using a limited set of images (45 persons, two views per person, with different facial expressions such as neutral vs. smiling), recognition performance as a function of the number of eigenvectors was measured for eigenfaces only and for the feature representations. For lower-order spaces, the eigenfeatures performed better than the eigenfaces (127); when the combined set was used, only marginal improvement was obtained. These experiments support the claim that feature-based mechanisms may be useful when gross variations are present in the input images.

It has been argued that practical systems should use a hybrid of PCA and local feature analysis (LFA). LFA is an interesting biologically inspired feature analysis method (126). Its biological motivation comes from the fact that though a huge array of receptors (more than six million cones) exist in the human retina, only a small fraction of them are active, corresponding to natural objects/signals that are statistically redundant. From the activity of these sparsely distributed receptors, the brain has to discover where and what objects are in the field of view and recover their attributes. Consequently, one expects to represent the natural objects/signals in a subspace of lower dimensionality by finding a suitable parameterization. For a limited class of objects such as faces, which are correctly aligned and scaled, this suggests that even lower dimensionality can be expected (126). One good example is the successful use of the truncated PCA expansion to approximate the frontal face images in a linear subspace (85).

Going a step further, the whole face region stimulates a full 2D array of receptors, each of which corresponds to a location in the face, but some of these receptors may be inactive. To explore this redundancy, LFA is used to extract topographic local features from the global PCA modes. Unlike PCA kernels Φ_i, which contain no topographic information (their supports extend over the entire grid of images), LFA kernels $K(\mathbf{x}_i, \mathbf{y})$ at selected grids $\mathbf{x}_i$ have local support. The search for the best topographic set of sparsely-distributed grids $\{x_o\}$ based on reconstruction error is called sparsification and is described in (126). Two interesting points are demonstrated in this work: (1) Using the same number of kernels, the perceptual reconstruction quality of LFA based on the optimal set of grids is better than that of PCA; the mean square error is 227, and 184 for a particular input and (2) keeping the second PCA eigenmodel in LFA reconstruction reduces the mean square error to 152, suggesting the hybrid use of PCA and LFA. No results on recognition performance based on LFA were reported.

A flexible appearance model-based method for automatic face recognition was presented in (92). To identify a face, both shape and gray-level information are modeled and used. The shape model is an active shape model; these are statistical models of the shapes of objects which iteratively deform to fit to an example of the shape in a new image. The statistical shape model is trained on example images using PCA, where the variables are the coordinates of the shape model points. For the purpose of classification, the shape variations due to interclass variation are separated from those due to within-class variations (such as small variations in 3D orientation and facial expression) using discriminant analysis. Based on the average shape of the shape model, a global shape-free gray-level model can be constructed, again using PCA. To further enhance the robustness of the system against changes in local appearance such as occlusions, local gray-level models are also built on the shape model points. Simple local profiles perpendicular to the shape boundary are used. Finally, for an input image, all three types of information, including extracted shape parameters, shape-free image parameters, and local profiles, are used to compute a Mahalanobis distance for classification as illustrated in Fig. 2.4. Based on training 10 and testing 13 images for each of 30 individuals, the classification rate was 92% for the ten normal testing images and 48% for the three difficult images.

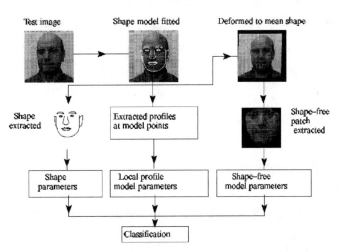

FIGURE 2.4: The face recognition scheme based on flexible appearance model (92). (Courtesy of A. Lanitis, C. Taylor, and T. Cootes.)

A component-based method for face detection/recognition was proposed in (66). The basic idea of component-based methods is to decompose a face into a set of facial components such as mouth and eyes that are interconnected by a flexible geometrical model. (This method is similar to the EBGM system (184) except that gray-scale components are used instead of Gabor wavelets). The motivation for using components is that changes in head pose mainly lead to changes in the positions of facial components, which could be accounted for by the flexibility of the geometric model. However, a major drawback of the system is that it needs a large number of training images taken from different viewpoints and under different lighting conditions.

2.2.8   Three-Dimensional Morphable Models

The method of using 3D morphable models for face recognition under situations of pose and illumination variations was proposed in (15). It overcomes the disadvantages of component-based methods by generating arbitrary synthetic images under varying pose and illumination, using a 3D morphable model. The 3D morphable model encodes shape and texture information in the model parameters. An algorithm is proposed that estimates these parameters from a single image. The shape and texture parameters are used to identify the individual.

The morphable face model is based on a vector space of faces. Any linear combination of shape and texture vectors from a set of example faces describes a realistic human face. Laser scans of 100 male and 100 female faces were recorded and stored in cylindrical coordinates relative to a vertical axis. The location of every 3D point is represented by three numbers, radius r, height h, and angular position ϕ. RGB values of face texture were also stored. The next step in building a morphable model is to obtain a dense point-to-point correspondence between each face and a reference face, which can be model from the database or any other 3D model. The dense correspondence, $v(h, \phi)$, is obtained by modifying optical flow for images to work for vector valued arrays, $I(h,]phi)$.

The coordinates and texture values of all n vertices in the reference face are concatenated to form shape and texture vectors $\mathbf{S}_0 = (x_1, y_1, z_1, \ldots, x_n, y_n, z_n)^T$

and $\mathbf{T}_0 = (R_1, G_1, B_1, \ldots, R_n, G_n, B_n)^T$. Using the optical flow field $v(h, \phi)$, vectors $\mathbf{S}_i$ and $\mathbf{T}_i$ are generated for each of the examples $i = 1, \ldots, M$ in the database. Any face, denoted as $(\mathbf{S}, \mathbf{T})$, can be represented using linear combination as

$$\mathbf{S} = \sum_{i=1}^{m} a_i \mathbf{S}_i \qquad \mathbf{T} = \sum_{i=1}^{m} b_i \mathbf{T}_I. \tag{2.7}$$

A PCA is performed separately on the set of shape and texture vectors. If $\{s_i\}$ and $\{t_i\}$ are the sets of eigenvectors corresponding to the shape and texture vectors and $\bar{s}$ and $\bar{t}$ are the corresponding mean values, then we have

$$\mathbf{S} = \bar{s} + \sum_{i=1}^{m-1} \alpha_i \mathbf{s}_i \qquad \mathbf{T} = \bar{t} + \sum_{i=1}^{m-1} \beta_i \mathbf{t}_i. \tag{2.8}$$

Given an input image $I_{\text{input}}(x, y)$, the goal is to estimate the model parameters α_I, β_i and face position, orientation, and illumination (represented by a vector ρ), such that the image rendered from the model is as close as possible to the input image. Assuming probability distributions on the parameters, this can be formulated as a maximum a-posteriori estimation problem, i.e. minimizing $E = -\ln(P(I_{\text{input}} \mid \alpha, \beta, \rho) p(\alpha, \beta, \rho))$. A stochastic gradient algorithm is used to obtain the optimum solution.

In the experiments, synthetic images of size 58×58 are generated for training both the detector and the classifier. Specifically, the faces were rotated in depth from $0°$ to $34°$ in $2°$ increments and rendered with two illumination models (the first model consists of ambient light alone and the second includes ambient light and a rotating point light source) at each pose. Fourteen facial components were used for face detection, but only nine components that were not strongly overlapped and contained gray-scale structures were used for classification. In addition, the face region was added to the nine components to form a single feature vector (a hybrid method), which was later trained by a SVM classifier (174). Training on three images and testing on 200 images per subject led to the following recognition rates on a set of six subjects: 90% for the hybrid method and roughly 10% for the global method that used the face region only; the false positive rate was 10%.

2.2.9 Illumination Effects in Face Recognition

As mentioned previously, many of the face recognition methods are affected by illumination variations. In (191), an analysis is carried out of how illumination variation changes the eigen-subspace projection coefficients of images under the assumption of a Lambertian surface. A varying-albedo Lambertian reflectance model, which relates the image I of an object to the object (p, q) (69), is used:

$$I = \rho \frac{1 + pP_s + qQ_s}{\sqrt{1 + p^2 + q^2}\sqrt{1 + P_s^2 + Q_s^2}} \tag{2.9}$$

where (p, q), ρ are the partial derivatives and varying albedo of the object, respectively. $(P_s, Q_s, -1)$ represents a single distant light source. The light source can also be represented by the illuminant slant and tilt angles; *slant* α is the angle between the opposite lighting direction and the positive z-axis, and *tilt* τ is the angle between the opposite lighting direction and the x–z plane. These angles are related to P_s and Q_s by $P_s = \tan\alpha \cos\tau$ and $Q_s = \tan\alpha \sin\tau$. To simplify the notation, we replace the constant $\sqrt{1 + P_s^2 + Q_s^2}$ by K. For easier analysis, it is assumed that frontal face objects are bilaterally symmetric about the vertical midlines of the faces.

Consider the basic expression for the subspace decomposition of a face image $I : I \simeq I_A + \sum_{i=1}^{m} a_i \Phi_i$, where I_A is the average image, Φ_i are the eigenimages, and a_i are the projection coefficients. Assume that for a particular individual we have a prototype image I_p that is a normally lighted frontal view [$P_s = 0$, $Q_s = 0$ in Eq. (2.9)] in the database, and we want to match it against a new image $\tilde{I}$ of the same class under lighting $(P_s, Q_s, -1)$. The corresponding subspace projection coefficient vectors $\mathbf{a} = [a_1, a_2, \ldots, a_m]^T$ (for I_p) and $\tilde{\mathbf{a}} = [\tilde{a}_1, \tilde{a}_2, \ldots, \tilde{a}_m]^T$ (for $\tilde{I}$) are computed as follows:

$$\begin{aligned} a_i &= I_p \odot \Phi_i - I_A \odot \Phi_i \\ \tilde{a}_i &= \tilde{I} \odot \Phi_i - I_A \odot \Phi_i, \end{aligned} \tag{2.10}$$

where $\odot$ denotes the sum of all element-wise products of two matrices (vectors). If we divide the images and the eigenimages into two halves, e.g. left and right, we have

$$a_i = I_P^L \odot \Phi_i^L + I_P^R \odot \Phi_i^R - I_A \odot \Phi_i$$
$$\tilde{a}_i = \tilde{I}^L \odot \Phi_i^L + \tilde{I}^R \odot \Phi_i^R - I_A \odot \Phi_i. \tag{2.11}$$

Based on (2.9), the symmetric property of eigenimages and face objects, we have

$$a_i = 2I_P^L[x, y] \odot \Phi_i^L[x, y] - I_A \odot \Phi_i$$
$$\tilde{a}_i = \left(\tfrac{2}{K}\right)\left(I_P^L[x, y] + I_P^L[x, y]q^L[x, y]Q_s\right) \odot \Phi_i^L[x, y] - I_A \odot \Phi_i, \tag{2.12}$$

leading to the following relation:

$$\tilde{\mathbf{a}} = \left(\frac{1}{K}\mathbf{a}\right) + \frac{Q_s}{K}[f_1^a, f_2^a, \ldots, f_m^a]^T - \frac{K-1}{K}\mathbf{a}_A. \tag{2.13}$$

where $f_i^a = 2(I_P^L[x, y]q^L[x, y]) \odot \Phi_i^L[x, y]$ and $\mathbf{a}_A$ is the projection coefficient vector of the average image $I_A : [I_A \odot \Phi_1, \ldots, I_A \odot \Phi_m]$. Now let us assume that the training set is extended to include mirror images as in (85). A similar analysis can be carried out, since in such a case the eigenimages are either symmetric (for most leading eigenimages) or antisymmetric. Thus (2.13) suggests that a significant illumination change can seriously degrade the performance of subspace-based methods.

In general, the illumination problem is quite difficult and has received considerable attention in the image understanding literature. In the case of face recognition, many approaches to this problem have been proposed that make use of the domain knowledge that all faces belong to one face class. These approaches can be divided into four types (191): (1) Heuristic methods, e.g. discarding the leading principal components; (2) image comparison methods in which appropriate image representations and distance measures are used; (3) class-based methods using multiple images of the same face in a fixed pose but under different lighting conditions; (4) model-based approaches in which 3D models are employed.

2.2.10 Model-Based Approaches

In model-based approaches, a 3D face model is used to synthesize the virtual image from a given image under desired illumination conditions. When the 3D model is unknown, recovering the shape from the images *accurately* is difficult without using any priors. Shape-from-shading (SFS) can be used if only one image is available; stereo or structure from motion (SfM) can be used when multiple images of the same object are available. Fortunately, for face recognition the differences in the 3D shapes of different face objects are not dramatic. This is especially true after the images are aligned and normalized. Recall that this assumption was used in the class-based methods reviewed above. Using a statistical representation of the 3D heads, PCA was suggested as a tool for solving the parametric SFS problem (3). An eigenhead approximation of a 3D head was obtained after training on about 300 laser-scanned range images of real human heads. The ill-posed SFS problem is thereby transformed into a parametric problem. The authors also demonstrate that such a representation helps to determine the light source. For a new face image, its 3D head can be approximated as a linear combination of eigenheads and then used to determine the light source. Using this complete 3D model, any virtual view of the face image can be generated. A major drawback of this approach is the assumption of *constant* albedo. This assumption does not hold for most real face images, even though it is the most common assumption used in SFS algorithms.

To address the issue of varying albedo, a *direct* 2D-to-2D approach was proposed based on the assumption that front-view faces are symmetric and making use of a generic 3D model (191). Recall that a prototype image I_p is a frontal view with $P_s = 0$, $Q_s = 0$. Substituting this into Eq. (2.9), we have

$$I_p[x, y] = \rho \frac{1}{\sqrt{1 + p^2 + q^2}}. \tag{2.14}$$

Comparing (2.9) and (2.14), we obtain

$$I_p[x, y] = \frac{K}{2(1 + q\, Q_s)}(I[x, y] + I[-x, y]). \tag{2.15}$$

This simple equation relates the prototype image I_p to $I[x, y] + I[x, -y]$, which is already available. The two advantages of this approach are (1) there is no need to recover the varying albedo $\rho[x, y]$ and (2) there is no need to recover the full shape gradients (p, q); q can be approximated by a value derived from a generic

3D shape. As part of the proposed automatic method, a model-based light source identification method was also proposed to improve existing source-from-shading algorithms. Using the Yale and Weizmann databases, noticeable performance improvements were reported when the prototype images are used in a subspace LDA system in place of the original input images (191). In these experiments, the gallery set contains about 500 images from various databases and the probe set contains 60 images from the Yale database and 96 images from the Weizmann database.

Recently, a general method of approximating Lambertian reflectance using second-order spherical harmonics has been reported (8). Assuming Lambertian objects under distant, isotropic lighting, the authors were able to show that the set of all reflectance functions can be approximated using the surface spherical harmonic expansion. Specifically, they have proved that using a second-order (nine harmonics, i.e. 9D-space) approximation, the accuracy for any light function exceeds 97.97%. They then extend this analysis to image formation, which is a much more difficult problem due to possible occlusion, shape, and albedo variations. As indicated by the authors, worst-case image approximation can be arbitrarily bad, but most cases are good. Using their method, an image can be decomposed into so-called *harmonic images*, which are produced when the object is illuminated by harmonic functions. The nine harmonic images of a face are plotted in Fig. 2.5.

Assuming precomputed object pose and known color albedo/texture, the authors reported an 86% correct recognition rate when applying this technique to the task of face recognition using a probe set of 10 people and a gallery set of 42 people.

2.2.11 Generalized Photometric Stereo

Generalized photometric stereo (198; 199) is an image-based approach we proposed to derive an illumination-invariant signature for face recognition. Same as ordinary photometric stereo algorithm (151), the generalized photometric stereo algorithm utilizes a Lambertian reflectance model to depict the visual appearance.

FIGURE 2.5: The first nine harmonic images of a face object (from left to right, top to bottom) (8). (Courtesy of R. Basri and D. Jacobs.)

In the Lambertian model, an image that is a collection of d pixels illuminated by the same light source $s_{3\times1}$ can be represented as

$$h_{d\times1} = T_{d\times3}\, s_{3\times1}, \tag{2.16}$$

where the T matrix encodes the products of albedo and unit surface normal vector for all pixels. The ordinary photometric stereo deals with n images of the *same* object, say $\{h_1, h_2, \ldots, h_n\}$, observed at a fixed pose illuminated by n different lighting sources, forming an *object-specific* ensemble. Simple algebraic manipulation gives

$$H_{d\times n} = [h_1, h_2, \ldots, h_n] = T_{d\times3}[s_1, s_2, \ldots, s_n]_{3\times n}. \tag{2.17}$$

Hence photometric stereo is rank-3 constrained. This requires capturing at least three images for one object in the gallery set, which might be prohibitive in practical scenarios.

The significant difference between ordinary and generalized photometric stereo (GPS) algorithms lies in the image ensemble they analyze. The image ensemble that the GPS algorithm analyzes consists of the appearances of different objects, with each object under a different illumination. However, both ordinary and GPS algorithms assume that all the images have been captured in a fixed view (here we used the frontal pose). To analyze the latter image ensemble, a rank constraint is proposed in (198; 199) on the product of albedo and surface normal. It assumes that any T matrix is a linear combination of some basis matrices $\{\mathsf{T}_1, \mathsf{T}_2, \ldots, \mathsf{T}_m\}$ coming from some m *basis objects*. Mathematically, there exist coefficients $\{f_j; \ j = 1, \ldots, m\}$ such that

$$\mathsf{T}_{d \times 3} = \sum_{j=1}^{m} f_j \mathsf{T}_j. \tag{2.18}$$

Substitution of (2.18) into (2.16) yields

$$\mathsf{h}_{d \times 1} = \mathsf{T}\mathsf{s} = \mathsf{W}_{d \times 3m}(\mathsf{f}_{m \times 1} \otimes \mathsf{s}_{3 \times 1}), \tag{2.19}$$

where $\mathsf{W} = [\mathsf{T}_1, \mathsf{T}_2, \ldots, \mathsf{T}_m]$, $\mathsf{f} = [f_1, f_2, \ldots, f_m]^\mathsf{T}$, and $\otimes$ denotes the Kronecker (tensor) product. This leads to a two-factor bilinear analysis (48). Notice that the W matrix encodes all albedos and surface normals for a class of objects.

With the availability of n images $\{\mathsf{h}_1, \mathsf{h}_2, \ldots, \mathsf{h}_n\}$ for *different* objects, observed at a fixed pose illuminated by n different lighting sources, forming a *class-specific* ensemble, we have

$$\mathsf{H}_{d \times n} = [\mathsf{h}_1, \mathsf{h}_2, \ldots, \mathsf{h}_n] = \mathsf{W}_{d \times 3m}[\mathsf{f}_1 \otimes \mathsf{s}_1, \mathsf{f}_2 \otimes \mathsf{s}_2, \ldots, \mathsf{f}_m \otimes \mathsf{s}_m]_{3m \times n}. \tag{2.20}$$

It is a rank-$3m$ factorization, which combines the rank of 3 for the illumination and the rank of m for the identity. A *class-specific* ensemble consists of exemplar images of different objects with each under a different illumination, which is beyond what can be analyzed using the bilinear analysis of (48). Bilinear analysis in (48) requires exemplar images of different objects under the same set of illuminations.

Because a factorization is always up to an invertible matrix, a full recovery of the albedos and surface normals is not a trivial task and requires additional constraints. We used two constraints: surface integrability and face symmetry. The surface integrability constraint (46; 47) has been used in several approaches (55; 188) to successfully recover albedo and shape. Suppose that the surface function is $z = z_{(x,y)}$, an integral surface must satisfy $\frac{\partial}{\partial x}\frac{\partial z}{\partial y} = \frac{\partial}{\partial y}\frac{\partial z}{\partial x}$. If given the unit surface normal vector $n_{(x,y)} \doteq [\hat{a}_{(x,y)}, \hat{b}_{(x,y)}, \hat{c}_{(x,y)}]^T$ at pixel (x, y), the integrability constraint requires that

$$\frac{\partial}{\partial x} \frac{\hat{b}_{(x,y)}}{\hat{c}_{(x,y)}} = \frac{\partial}{\partial y} \frac{\hat{a}_{(x,y)}}{\hat{c}_{(x,y)}}. \qquad (2.21)$$

The symmetry constraint has also been employed in (152; 192) for face images. For a face image in a frontal view, it is symmetric about the central y-axis (152; 192):

$$p_{(x,y)} = p_{(-x,y)}; \hat{a}_{(x,y)} = -\hat{a}_{(-x,y)}; \hat{b}_{(x,y)} = \hat{b}_{(-x,y)}; \hat{c}_{(x,y)} = \hat{c}_{(-x,y)}, \qquad (2.22)$$

where $p_{(x,y)}$ is the albedo for the pixel (x, y).

In order to fuse the constraints to recover the *class-specific* albedos and surface normals (or equivalently the W matrix), with no knowledge of the light sources and in the presence of shadows, an optimization approach was taken in (198; 199). More importantly, this approach takes into account the effects of a varying albedo field by approximating the integrability terms using only the surface normals instead of the product of the albedos and the surface normals. Because of the nonlinearity embedded in the integrability terms, regular algorithms such as the steepest descent are inefficient. Invoking vector derivatives, we derived a "linearized" algorithm to find the solution. Fig. 2.6 illustrates how the generalized photometric stereo algorithm works.

The GPS algorithm was evaluated for a face recognition application. The matrix W was first estimated from a given training set and subsequently used to cope with arbitrary images belonging to individuals in the gallery and probe sets that do not overlap with the training set in terms of the identity. Since the blending coefficient provides an identity encoding that is invariant to illumination, it is employed for face recognition under illumination variation, which results in good recognition performance. Using the PIE database (154), an average of 93% that compares favorably to other state-of-the-art approaches. Also two byproducts of the above process are image synthesis and illumination estimation.

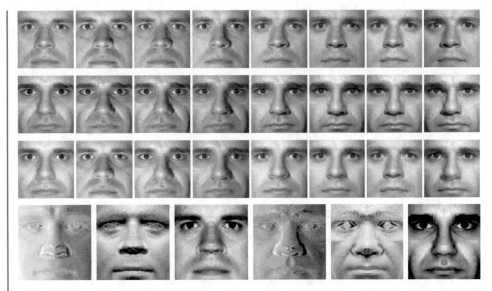

FIGURE 2.6: (Row 1) The first basis object under eight different illuminations. (Row 2) The second basis object under the same set of eight different illuminations. (Row 3) Eight images (constructed by random linear combinations of two basis objects) illuminated by eight different lighting sources. (Row 4) Recovered class-specific albedo-shape matrix W showing the product of varying albedos and surface normals of two basis objects (i.e. the three columns of T_1 and T_2) using the generalized photometric stereo algorithm.

2.2.12 Illuminating Light Field

The generalized photometric stereo approach handles only illumination variation not pose variation, another major source of variation affecting face recognition performance. In (195; 196), the GPS approach was extended to a so-called illuminating light field approach to deal with pose variation as well. Consider a set of fixed poses $\{v_1, v_2, \ldots, v_K\}$. The corresponding images at these poses are $\{h^{v_1}, h^{v_2}, \ldots, h^{v_K}\}$. Principally, starting from these K images, one is able to construct a light field (97) that measures the radiance in free space (free of occluders) as a 4D function of position and direction to characterize the continuous pose space.

For the sake of simplicity, formulate a long $Kd \times 1$ vector L^s by stacking all the images (each image is a $d \times 1$ vector) under the illumination s at all these poses, $L^s = [h^{v_1 s\,T}, h^{v_2 s\,T}, \ldots, h^{v_K s\,T}]^T$. Since all these images are illuminated by the same light source s, the GPS approach can be used to perform a bilinear analysis:

$$\mathsf{L}^s_{Kd\times1} = \mathsf{W}_{Kd\times3m}(\mathsf{f}_{m\times1} \otimes \mathsf{s}_{3\times1}).$$ (2.23)

An image h^{vs} under the pose v and illumination s is

$$\mathsf{h}^{vs}_{d\times1} = \mathcal{R}_v[\mathsf{L}^s] = \mathcal{R}_v[\mathsf{W}(\mathsf{f} \otimes \mathsf{s})] = \mathsf{W}^v_{d\times3m}(\mathsf{f}_{m\times1} \otimes \mathsf{s}_{3\times1}),$$ (2.24)

where $\mathcal{R}_v$ is a row sampling operator and $\mathsf{W}^v = \mathcal{R}_v[\mathsf{W}]$. Equation (2.24) has an important implication: The coefficient vector f provides an identity signature invariant to both pose and illumination because the pose is absorbed in W^v and the illumination is absorbed in s. Using this invariant signature, face recognition experiments using the PIE database (154) were performed. Significant improvement over other subspace methods was reported in (196). However, the 3D morphable model approach recorded better performance when dealing with pose variation. Detailed comparisons between the illuminating light field approach and the 3D morphable model approach are highlighted in (196).

2.2.13 Summary of Image-Based Face Recognition

Face recognition continues to be a very active area of research. It continues to adopt state-of-the-art techniques from learning, computer vision, and pattern recognition. Among all face detection/recognition methods, appearance/image-based approaches seem to have dominated up to now. The main reason is the strong prior that all face images belong to a face class. However, 3D model-based approaches will probably become more popular as face recognition methodologies make use of video sequences. Hybrid face recognition systems that use both holistic and local features resemble the human perceptual system. While the holistic approach provides a quick recognition method, the discriminant information that it provides may not be rich enough to handle very large databases. This insufficiency can be compensated for by local feature methods. However, issues remain in accurately locating facial features and deciding on the importance of holistic and local features. At a broad level, occlusion, poor image quality due to varying lighting conditions (especially outdoors), pose variations and aging continue to be the main challenges facing researchers in this area.

2.3   FACE RECOGNITION FROM VIDEO

In surveillance, information security, and access control applications, face recognition and identification from a video sequence is an important problem. Face recognition based on video is preferable over using still images, since motion helps in recognition of (familiar) faces (87). It has also been demonstrated that humans can recognize animated faces better than randomly rearranged images from the same set. The problems facing video-based face recognition are (a) poor quality of video (especially outdoors) and the large changes in illumination and pose and (b) detecting and localizing a face, which is often very small, from the background clutter, which may vary from frame to frame. We begin our description of this problem with a broad overview of some basic techniques, briefly review some of the more well-known ones and then describe one of the methods we have developed recently in relative detail.

2.3.1   Basic Techniques of Video-Based Face Recognition

Based on the current state of the art in video-based face recognition, we review three specific face-related techniques: face segmentation and pose estimation, face tracking, and face modeling.

Face Segmentation and Pose Estimation: Some of the techniques that have been used for face segmentation are pixel-based change detection using difference images, localization using motion and color information, and different heuristics using a combination of the above. Given a face region, important facial features can be located. The locations of feature points can be used for pose estimation which is important for synthesizing a virtual frontal view. Newly developed segmentation methods locate the face and estimate its pose simultaneously without extracting features (62; 100).

Face Tracking: Face and feature tracking are critical for reconstructing a face model, facial expression recognition, and gaze recognition. In its most general form, tracking is essentially motion estimation. However, face tracking can take advantage of domain knowledge (e.g. use of 2D or 3D models) and can be divided into three parts: head tracking, facial feature tracking, and a combination of both. Early efforts focused on the first two problems: head tracking (5) and facial feature tracking (168; 189). In (13), a tracking system based on local parameterized models is used to recognize facial expressions. The use of 3D face models and SfM for tracking was proposed in (74; 161). Some of the newest model-based tracking methods calculate the 3D motions and deformations directly from image intensities (20), thus eliminating the information-lossy intermediate representations.

Face Modeling: Modeling of faces includes 3D shape modeling and texture modeling. In computer vision, one of the most widely used methods of estimating 3D shape from a video sequence is SfM, which estimates the 3D depths of interesting points. The unconstrained SfM problem has been approached in two ways. In the differential approach, one computes some type of flow field (optical, image, or normal) and uses it to estimate the depths of visible points. The difficulty in this approach is reliable computation of the flow field. In the discrete approach, a set of features such as points, edges, corners, lines, or contours are tracked over a sequence of frames and the depths of these features are computed.

Various researchers have addressed the issue of 3D face modeling. In (75), the authors used an extended Kalman filter to recover the 3D structure of a face, which was then used for tracking. A method for recovering nonrigid 3D shapes as a linear combination of a set of basis shapes was proposed in (21). A factorization based method for recovering nonrigid 3D structure and motion from video was presented in (19). In (133), the author proposes a method for self-calibration in the presence of varying internal camera parameters and reconstructs metric 3D structure. Romdhani *et al.* (138) have shown that it is possible to recover the shape and texture parameters of a 3D Morphable Model from a single image. They used their 3D model for identification of faces under different pose and illumination conditions. Three-dimensional face modeling using bundle adjustment was proposed in (49) and (149). Their method initialized the reconstruction algorithm with a generic model. A method using SfM for face reconstruction without biasing the model toward a generic face was recently proposed in (145). The use of 3D models for face recognition will be explored further in (145). This issue of 3D models for face recognition will be discussed in greater detail in the chapter on future research directions.

2.3.2   Overview of Various Methods in Video-Based Face Recognition

Historically, video face recognition originated from still-image based techniques. That is, the system automatically detects and segments the face from the video, and then applies still-image face recognition techniques, similar to the ones described in Section 2.2. An improvement over these methods is to apply tracking; this can help in recognition, in that a virtual frontal view can be synthesized via pose and depth estimation from video. Because of the abundance of frames in a video, another way to improve the recognition rate is the use of "voting" based on the recognition results from each frame (57).

Several practical systems have been developed that take a video sequence as the input and try to identify the face in the sequence. In many applications, multimodal recognition schemes that rely on face, body motion, fingerprint, speech, etc. have also been developed. However, in our opinion, most of the existing practical systems do not fully exploit the video sequence.

In (181), a fully automatic person authentication system is described which includes video break, face detection, and authentication modules. The video break module, corresponding to key-frame detection based on object motion, consisted of two units. The first unit implemented a simple optical flow method; it was used when the image SNR level was low. When the SNR level was high, simple pairwise frame differencing was used to detect the moving object. The face detection module consisted of three units: face localization using analysis of projections along the x- and y-axes; face region labeling using a decision tree learned from positive and negative examples taken from 12 images each consisting of 2759 windows of size 8×8; face normalization based on the numbers of face region labels. The normalized face images are then used for authentication, using a radial basis function (RBF) network. This system was tested on three image sequences; the first was taken indoors with one subject present, the second was taken outdoors with two subjects, and the third was taken outdoors with one subject under stormy conditions. Perfect results were reported on all three sequences, as verified against a database of 20 still face images.

An access control system based on person authentication is described in (109). The system combines two complementary visual cues: motion and facial appearance. In order to reliably detect significant motion, spatio-temporal zero crossings computed from six consecutive frames were used. These motions were grouped into moving objects using a clustering algorithm, and Kalman filters were employed to track the grouped objects. An appearance-based face detection scheme using RBF networks was used to confirm that an object is a person. The face detection scheme was "bootstrapped"using motion and object detection to provide an approximate head region. Face tracking based on the RBF network was used to provide feedback to the motion clustering process to help deal with occlusions. Good tracking results were demonstrated. The method was then used for person authentication using PCA or LDA.

An appearance model based method for video tracking and enhancing identification is proposed in (42). The appearance model is a combination of the active shape model (32) and the shape-free texture model after warping the face into a mean shape. The authors use a combined set of parameters for both models. The main contribution is the decomposition of the combined model parameters into an identity subspace and an orthogonal residual subspace using linear discriminant analysis. The residual subspace ideally contains intraperson variations that are caused by pose, lighting, and expression. Examples of face tracking and visual enhancement are demonstrated, but no recognition experiments are reported.

In (159), a system called PersonSpotter is described. This system is able to capture, track, and recognize a person walking toward or passing a stereo CCD camera. It has several modules, including a head tracker, preselector, landmark finder, and identifier. An elastic graph matching scheme is employed to identify the face. A recognition rate of about 90% was achieved; the size of the database is not known.

A multimodal person recognition system is described in (28). This system consists of a face recognition module, a speaker identification module, and a classifier fusion module. It has the following characteristics: (1) The face recognition module can detect and compensate for pose variations and the speaker identification module can detect and compensate for changes in the auditory background; (2) the most reliable video frames and audio clips are selected for recognition; (3) 3D information about the head obtained through SfM is used to detect the presence of an actual person as opposed to an image of that person.

In (98), a face verification system based on tracking facial features is presented. The basic idea of this approach is to exploit the temporal information available in a video sequence to improve face recognition. First, the feature points defined by Gabor attributes on a regular 2D grid are tracked. Then, the trajectories of these tracked feature points are exploited to identify the person presented in a short video sequence. The proposed tracking-for-verification scheme is different from the pure tracking scheme in that one template face from a database of known persons is selected for tracking. For each template with a specific personal ID, tracking can be performed and trajectories can be obtained. Based on the characteristics of these trajectories, identification is carried out. According to the authors, the trajectories of the same person are more coherent than those of different persons. Such characteristics can also be observed in the posterior probabilities over time by assuming different classes. In other words, the posterior probabilities for the true hypothesis tend to be higher than those for false hypotheses. This in turn can be used for identification. Testing results on a small databases of 19 individuals suggest that performance is favorable over a frame-to-frame matching and voting scheme, especially in the case of large lighting changes. The testing result is based on comparison with alternative hypotheses.

In (99), a multiview-based face recognition system is proposed to recognize faces from videos with large pose variations. To address the challenging pose issue, the concept of an *identity surface* that captures joint spatial and temporal information is used. An identity surface is a hypersurface formed by projecting all the images of one individual onto the discriminating feature space parameterized with respect to head pose. To characterize the head pose, two angles, yaw and tilt are used as basis coordinates in the feature space. The other basis coordinates represent discriminating feature patterns of faces. Based on recovered pose information, a trajectory of the input feature pattern can be constructed. The trajectories of features from known subjects arranged in the same temporal order can be synthesized on their respective identity surfaces. To recognize a face across views over time, the trajectory for the input face is matched to the trajectories synthesized for the known subjects. This approach can be thought of as a generalized version of face recognition based on single images taken at different poses. Experimental results using 12 training sequences, each containing one subject, and new testing sequences of these subjects were reported. Recognition rates were 100% and 93.9%, using 10 and 2 KDA (kernel discriminant analysis) vectors, respectively.

2.3.3   Probabilistic Approaches in Video-Based Face Recognition

While most face recognition algorithms take still images as probe inputs, a video-based face recognition approach that takes video sequences as inputs has recently been developed (200). Since the detected face might be moving in the video sequence, one has to deal with uncertainty in tracking as well as in recognition. Rather than resolving these two uncertainties separately, (200) performs simultaneous tracking and recognition of human faces from a video sequence. We will now explain this methodology in detail.

In general, a video sequence is a collection of still images; so still-image-based recognition algorithms can always be applied. An important property of a video sequence, however, is its temporal continuity. While this property has been exploited for tracking, it has not been used for recognition. Here we systematically investigate how temporal continuity can be incorporated for video-based recognition.

Our probabilistic approach solves *still-to-video* recognition, where the gallery consists of still images and the probes are video sequences. A time series state space model is proposed to fuse temporal information in a probe video, which simultaneously characterizes the kinematics and identity using a *motion vector* θ_t and an *identity variable* n_t, respectively.

The recognition model consists of the following components:

1. *Motion equation:* In its most general form, the motion model can be written as

$$\theta_t = g(\theta_{t-1}, u_t), \qquad t \geq 1, \tag{2.25}$$

where u_t is *noise* in the motion model, whose distribution determines the motion state transition probability $p(\theta_t \mid \theta_{t-1})$. The function $g(., .)$ characterizes the evolving motion and it could be a function learned offline or given a priori. One of the simplest choice is an additive function, i.e., $\theta_t = \theta_{t-1} + u_t$, which leads to a first-order Markov chain.

2. *Identity equation:* Assuming that the identity does not change as time proceeds, we have

$$n_t = n_{t-1}, \qquad t \geq 1. \tag{2.26}$$

In practice, one may assume a small transition probability between identity variables to increase the robustness.

3. *Observation equation:* By assuming that the transformed observation is a noise-corrupted version of some still template in the gallery, the observation equation can be written as

$$\mathcal{T}_{\theta_t}\{z_t\} = I_{n_t} + v_t, \qquad t \geq 1, \tag{2.27}$$

where v_t is *observation noise* at time t, whose distribution determines the observation likelihood $p(z_t \mid n_t, \theta_t)$, I_{n_t} is the corresponding image for person n_t in the gallery set, and $T_{\theta_t}\{z_t\}$ is a transformed version of the observation z_t. This transformation could be either geometric or photometric or both.

4. *Statistical independence:* Statistical independence between all noise variables u_t's and v_t's is assumed.

5. *Prior distribution:* The prior distribution $p(n_0 \mid z_0)$ is assumed to be uniform, i.e.,

$$p(n_0 \mid z_0) = N^{-1}; \qquad n_0 = 1, 2, \ldots, N. \tag{2.28}$$

In our experiments, $p(\theta_0 \mid z_0)$ is assumed to be Gaussian, whose mean comes from an initial detector or a manual input and whose covariance matrix is manually specified.

Using time recursion, Markov properties, and statistical independence embedded in the model, one can derive

$$
\begin{aligned}
p(n_{0:t}, \theta_{0:t} \mid z_{0:t}) &= p(n_{0:t-1}, \theta_{0:t-1} \mid z_{0:t-1}) \frac{p(z_t \mid n_t, \theta_t) p(n_t \mid n_{t-1}) p(\theta_t \mid \theta_{t-1})}{p(z_t \mid z_{0:t-1})} \\
&= p(n_0, \theta_0 \mid z_0) \prod_{s=1}^{t} \frac{p(z_s \mid n_s, \theta_s) p(n_s \mid n_{s-1}) p(\theta_s \mid \theta_{s-1})}{p(z_s \mid z_{0:s-1})} \\
&= p(n_0 \mid z_0) p(\theta_0 \mid z_0) \prod_{s=1}^{t} \frac{p(z_s \mid n_s, \theta_s) \delta(n_s - n_{s-1}) p(\theta_s \mid \theta_{s-1})}{p(z_s \mid z_{0:s-1})}.
\end{aligned}
\tag{2.29}
$$

Therefore, by marginalizing over $\theta_{0:t}$ and $n_{0:t-1}$, we obtain

$$p(n_t = l \mid z_{0:t}) = p(l \mid z_0) \int_{\theta_0} \cdots \int_{\theta_t} p(\theta_0 \mid z_0) \prod_{s=1}^{t} \frac{p(z_s \mid l, \theta_s) p(\theta_s \mid \theta_{s-1})}{p(z_s \mid z_{0:s-1})} d\theta_t \ldots d\theta_0. \tag{2.30}$$

Thus $p(n_t = l \mid z_{0:t})$ is determined by the prior distribution $p(n_0 = l \mid z_0)$ and the product of the likelihood functions, $\prod_{s=1}^{t} p(z_s \mid l, \theta_s)$. If a uniform prior is assumed, then $\prod_{s=1}^{t} p(z_s \mid l, \theta_s)$ is the only determining factor that accumulates the evidence.

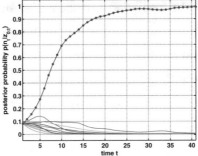

FIGURE 2.7: (Top row) The gallery images. (Bottom row) The first (left) and the last (middle) frames of the video sequences with tracking results indicated by the box and the posterior probability $p(n_t \mid z_{0:t})$.

As this model is nonlinear with non-gaussian noise, the joint posterior distribution of the motion vector and the identity variable is approximated at each time instant using a computationally efficient *sequential importance sampling* (SIS) algorithm and then propagated to the next time instant. *Marginalization* over the motion vector yields a robust estimate of the posterior distribution of the identity variable. Empirical results as in Fig. 2.7 demonstrate that, due to the propagation of the identity variable over time, a *degeneracy* in posterior probability of the identity variable is achieved to obtain improved recognition.

Recently, an algorithm was presented for recognizing human faces from video sequences with significant pose variations (94). Each registered person is represented by a low dimensional appearance manifold in the image space. The complex nonlinear appearance manifold is expressed as a collection of subsets (called pose manifolds) and the connectivity among them. Each pose manifold is approximated by an affine plane. In order to construct this representation, exemplars are sampled from videos and then clustered with a K-means algorithm. Each cluster is represented as a plane computed through principal components analysis. The connectivity between the pose manifolds encodes the transition probability between images in each of the pose manifolds and is learned from a training video sequence. Face recognition from a test video sequence is achieved using a maximum *a posteriori* estimation, by integrating the likelihood that the input image comes from a particular pose manifold and the transition probability to this pose manifold from the previous frame. Defining overall recognition rate by dividing the number of frames where the identity is correctly recognized by the total number of frames in all the test videos, the authors reported a recognition rate of over 90% on a database of 20 people.

2.4 EVALUATION OF FACE RECOGNITION TECHNOLOGIES

Given the numerous theories and techniques that are applicable to face recognition, it is clear that evaluation and benchmarking of these algorithms is crucial. Until recently, there did not exist a common FRT evaluation protocol that included large databases and standard evaluation methods. This made it difficult to assess the status of FRT for real applications, even though many existing systems reported almost perfect performance on small databases. In this section, we will describe the methodology of and the results from the most extensive evaluation of FRT technologies, the FERET evaluation.

2.4.1 The FERET Evaluation

The first FERET evaluation test was administered in August 1994 (132). This evaluation established a baseline for face recognition algorithms and was designed to measure performance of algorithms that could automatically locate, normalize, and identify faces. This evaluation consisted of three tests, each with a different gallery and probe set. (A gallery is a set of known individuals, while a probe is a set of unknown faces presented for recognition.) The first test measured identification performance from a gallery of 316 individuals with one image per person, the second was a false-alarm test, and the third measured the effects of pose changes on performance. The second FERET evaluation was administered in March 1995; it consisted of a single test that measured identification performance from a gallery of 817 individuals, and included 463 duplicates in the probe set (132). (A duplicate is a probe for which the corresponding gallery image was taken on a different day; there were only 60 duplicates in the Aug94 evaluation.) The third and last evaluation (Sep96) was administered in September 1996 and March 1997.

Database: The images were collected in 15 sessions between August 1993 and July 1996. Each session lasted 1 or 2 days, and the location and setup did not change during the session. Sets of 5–11 images of each individual were acquired under relatively unconstrained conditions (see Fig. 2.8). They included two frontal views; in the first of these (**fa**) a neutral facial expression was requested and in the second (**fb**) a different facial expression was requested (these requests were

FIGURE 2.8: Images from the FERET data set; these images are of size 384 × 256.

not always honored). For 200 individuals, a third frontal view was taken using a different camera and different lighting; this is referred to as the **fc** image. The remaining images were nonfrontal and included right and left profiles, right and left quarter profiles, and right and left half profiles. The FERET database consists of 1564 sets of images (1199 original sets and 365 duplicate sets)—a total of 14 126 images. A development set of 503 sets of images was released to researchers; the remaining images were sequestered for independent evaluation. In late 2000 the entire FERET database was released along with the Sep96 evaluation protocols, evaluation scoring code, and baseline PCA algorithms.

Evaluation: For details of the three FERET evaluations see (131; 132). The results of the Sep96 FERET evaluation will be briefly reviewed here. Because the entire FERET data set has been released, the Sep96 protocol provides a good benchmark for performance of new algorithms. For the Sep96 evaluation, there was a primary gallery consisting of one frontal image (**fa**) per person for 1196 individuals. This was the core gallery used to measure performance for the following four different probe sets.

- **fb** probes: Gallery and probe images of an individual taken on the same day with the same lighting (1195 probes).

- **fc** probes: Gallery and probe images of an individual taken on the same day with different lighting (194 probes).

- Dup I probes: Gallery and probe images of an individual taken on different days—duplicate images (722 probes).

- Dup II probes: Gallery and probe images of an individual taken over a year apart (the gallery consisted of 894 images; 234 probes).

Performance was measured using two basic methods. The first measured identification performance, where the primary performance statistic is the percentage of probes that are correctly identified by the algorithm. The second measured verification performance, where the primary performance measure is the equal error rate between the probability of false alarm and the probability of correct verification. (A more complete method of reporting identification performance is a cumulative match characteristic; for verification performance it is a receiver operating characteristic (ROC).)

The Sep96 evaluation tested the following ten algorithms:

- An algorithm from Excalibur Corporation (Carlsbad, CA)(Sept. 1996)
- Two algorithms from MIT Media Laboratory (Sept. 1996) (113; 173)
- Three LDA-based algorithms from Michigan State University (166) (Sept. 1996) and the University of Maryland (44; 193) (Sept. 1996 and March 1997)
- A gray-scale projection algorithm from Rutgers University (182) (Sept. 1996)
- An elastic graph matching algorithm from the University of Southern California (184) (March 1997)
- A baseline PCA algorithm (173)
- A baseline normalized correlation matching algorithm.

Three of the algorithms performed very well: probabilistic eigenface from MIT (113), subspace LDA from UMD (191; 193), and elastic graph matching from USC (184).

A number of lessons were learned from the FERET evaluations. The first is that performance depends on the probe category and there is a difference between best and average algorithm performance.

Another lesson is that the scenario has an impact on performance. For identification, on the **fb** and duplicate probes, the USC scores were 94% and 59%, and the UMD scores were 96% and 47%. However, for verification, the equal error rates were 2% and 14% for USC and 1% and 12% for UMD.

Another important contribution of the FERET evaluations is the identification of areas for future research. In general the test results revealed three major problem areas: recognizing duplicates, recognizing people under illumination variations, and under pose variations.

2.4.2   Face Recognition Vendor Test 2000

The Sep96 FERET evaluation measured performance on prototype laboratory systems. After March 1997 there was rapid advancement in the development of commercial face recognition systems. This advancement represented both a maturing of face recognition technology, and the development of the supporting system and infrastructure necessary to create commercial off-the-shelf (COTS) systems. By the beginning of 2000, COTS face recognition systems were readily available.

To assess the state of the art in COTS face recognition systems the FRVT 2000 was organized (14). FRVT 2000 was a technology evaluation that used the Sep96 evaluation protocol, but was significantly more demanding than the Sep96 FERET evaluation.

Participation in FRVT 2000 was restricted to COTS systems, with companies from Australia, Germany, and the United States participating. The five companies evaluated were Banque-Tec International Pty. Ltd., C-VIS Computer Vision und Automation GmbH, Miros, Inc., Lau Technologies, and Visionics Corporation.

A greater variety of imagery was used in FRVT 2000 than in the FERET evaluations. FRVT 2000 reported results in eight general categories: compression, distance, expression, illumination, media, pose, resolution, and temporal. There was no common gallery across all eight categories; the sizes of the galleries and probe sets varied from category to category.

We briefly summarize the results of FRVT 2000. Complete details can be found in (14), and include identification and verification performance statistics. The media experiments showed that changes in media do not adversely affect performance. Images of a person were taken simultaneously on conventional film and on digital media. The compression experiments showed that compression does not adversely affect performance. Probe images compressed up to 40:1 did not reduce recognition rates. The compression algorithm was JPEG.

FRVT 2000 also examined the effect of pose angle on performance. The results show that pose does not significantly affect performance up to $\pm 25°$, but that performance is significantly affected when the pose angle reaches $\pm 40°$.

In the illumination category, two key effects were investigated. The first was lighting change indoors. This was equivalent to the **fc** probes in FERET. For the best system in this category, the indoor change of lighting did not significantly affect performance. A second experiment tested recognition with an indoor gallery and an outdoor probe set. Moving from indoor to outdoor lighting significantly affected performance, with the best system achieving an identification rate of only 0.55.

The temporal category is equivalent to the duplicate probes in FERET. To compare progress since FERET, dup I and dup II scores were reported. For FRVT 2000 the dup I identification rate was 0.63 compared with 0.58 for FERET. The corresponding rates for dup II were 0.64 for FRVT 2000 and 0.52 for FERET. These results show that there was algorithmic progress between the FERET and FRVT 2000 evaluations. FRVT 2000 showed that two common concerns, the effects of compression and recording media, do not affect performance. It also showed that future areas of interest continue to be duplicates, pose variations, and illumination variations generated when comparing indoor images with outdoor images.

2.4.3   Face Recognition Vendor Test 2002

The FRVT 2002 (130) was a large-scale evaluation of automatic face recognition technology. The primary objective of FRVT 2002 was to provide performance measures for assessing the ability of automatic face recognition systems to meet real-world requirements. Ten participants were evaluated under the direct supervision of the FRVT 2002 organizers in July and August 2002.

The heart of the FRVT 2002 was the high computational intensity test (HCInt). The HCInt consisted of 121 589 operational images of 37 437 people.

The images were provided from the U.S. Department of State's Mexican nonimmigrant Visa archive. From this data, real-world performance figures on a very large data set were computed. Performance statistics were computed for verification, identification, and watch list tasks.

FRVT 2002 results show that normal changes in indoor lighting do not significantly affect performance of the top systems. Approximately the same performance results were obtained using two indoor data sets, with different lighting, in FRVT 2002. In both experiments, the best performer had a 90% verification rate at a false accept rate of 1%. On comparable experiments conducted 2 years earlier in FRVT 2000, the results of FRVT 2002 indicate that there has been a 50% reduction in error rates. For the best face recognition systems, the recognition rate for faces captured outdoors, at a false accept rate of 1%, was only 50%. Thus, face recognition from outdoor imagery remains a research challenge area.

A very important question for real-world applications is the rate of decrease in performance as time increases between the acquisition of the database of images and new images presented to a system. FRVT 2002 found that for the top systems, performance degraded at approximately 5% per year.

One open question in face recognition is: How does database and watch list size effect performance? Because of the large number of people and images in the FRVT 2002 data set, FRVT 2002 reported the first large-scale results on this question. For the best system, the top-rank identification rate was 85% on a database of 800 people, 83% on a database of 1,600, and 73% on a database of 37 437. For every doubling of database size, performance decreases by two to three overall percentage points. More generally, identification performance decreases linearly in the logarithm of the database size.

Previous evaluations have reported face recognition performance as a function of imaging properties. For example, previous reports compared the differences in performance when using indoor vs. outdoor images, or frontal vs. nonfrontal images. FRVT 2002, for the first time, examined the effects of demographics on performance. Two major effects were found. First, recognition rates for males were higher than females. For the top systems, identification rates for males were 6–9% points higher than that of females. For the best system, identification performance on males was 78% and for females it was 79%. Second, recognition rates for older people were higher than for younger people. For 18- to 22-year olds the average identification rate for the top systems was 62%, and for 38- to 42-year olds it was 74%. For every 10-year increase in age, performance increases on the average by approximately 5% through age 63.

FRVT 2002 looked at two of these new techniques. The first was the 3D morphable models technique of Blanz and Vetter (15). Morphable models provide a technique for improving recognition of nonfrontal images. FRVT 2002 found that Blanz and Vetter's technique significantly increased the recognition performance. The second technique is recognition from video sequences. Using FRVT 2002 data, it was found that recognition performance using video sequences offered only a limited improvement in performance over still images.

In summary, the key lessons learned in FRVT 2002 are as follows: (1) Given reasonable controlled indoor lighting, the current state of the art in face recognition is 90% verification at a 1% false accept rate. (2) Face recognition in outdoor images is an unsolved problem. (3) The use of morphable models can significantly improve nonfrontal face recognition. (4) Identification performance decreases linearly in the logarithm of the size of the gallery. (5) In face recognition applications, accommodations should be made for including demographic information since characteristics such as age and sex can significantly affect performance.

2.4.4   The XM2VTS Protocol

Multimodal methods[1] are a very promising approach to user-friendly (hence acceptable), highly secure personal verification. Recognition and verification systems need training; the larger the training set, the better the performance achieved. The volume of data required for training a multimodal system based on analysis of video

---

[1] http://www.ee.surrey.ac.uk/Research/VSSP/xm2vtsdb/.

and audio signals is on the order of TBytes; technology that allows manipulation and effective use of such volumes of data has only recently become available in the form of digital video. The XM2VTS multimodal database (111) contains four recordings of 295 subjects taken over a period of 4 months. Each recording contains a speaking head shot and a rotating head shot. Available data from this database include high-quality color images, 32 KHz 16-bit sound files, video sequences, and a 3D model.

2.5   CONCLUSIONS

In this chapter, we have provided a broad overview of the various methods that have been applied to face recognition. We divided the various methods into two categories: recognition from still images and recognition from video. We described some of the more well-known methods in some detail: eigenfaces, elastic bunch graph matching, linear discriminant analysis or Fisherfaces with single and multiple exemplars, morphable models, and probabilistic models for recognition from video. In addition, we dealt with some of the issues that affect face recognition, namely pose and illumination variation. For illumination variation, we analyzed its effect on the eigen-subspace projection coefficients of images and discussed a method based on the GPS techniques. For pose variations, we described a method for building 3D face models using shape-from-shading and a hybrid method that combines GPS and light-fields. The use of 3D models will be discussed in greater detail in a later chapter. We will now summarize the main conclusions that can be drawn based on our experience on working in the area of face recognition:

- Machine recognition of faces has emerged as an active research area spanning disciplines such as image processing, pattern recognition, computer vision, and neuroscience. There are numerous applications of FRT to commercial systems such as face verification based ATM and access control, as well as law enforcement applications to video surveillance, etc.

- Extensive research in psychophysics and the neurosciences on human recognition of faces is documented in the literature. It is beneficial for engineers who design face recognition systems to be aware of the relevant findings.

- Preprocessing steps, like face detection, segmentation, feature localization, tracking, are key to robust and accurate face recognition. During the past several years, significant achievements have been made in this area. However, the difficulty of these tasks under any image acquiring condition partially explains why robust face recognition is still a difficult problem.

- Numerous methods have been proposed for face recognition based on image intensities. Many of these methods have been successfully applied to the task of face recognition, but they have advantages and disadvantages. The choice of a method should be based on the specific requirements of a given task. For example, the EBGM-based method (184) has very good performance, but it requires an image size, e.g., 128×128, which severely restricts its possible application to video-based surveillance where the image size of the face area is very small. On the other hand, the subspace LDA method (191) works well for both large and small images, e.g., 96×84 or 12×11.

- Recognition of faces from a video sequence (especially, a surveillance video) is still one of the most challenging problems in face recognition because video is of low quality and the images are small. Often, the subjects of interest are not cooperative, e.g., not looking into the camera. Nevertheless, video-based face recognition systems using multiple cues have demonstrated good results in relatively controlled environments.

- A crucial step in face recognition is the evaluation and benchmarking of algorithms. One of the most important face databases (FERET) and its associated evaluation methods, the conclusions of FRVT 2000 and 2002, were reviewed. The availability of these evaluations has had a significant impact on progress in the development of face recognition algorithms.

- Although many face recognition techniques have been proposed and have shown significant promise, robust face recognition is still difficult. There are many major challenges: pose, illumination, aging and recognition in outdoor imagery. Some basic problems remain to be solved; for example, pose discrimination is not difficult but accurate pose estimation is hard. In addition to these two problems, there are other even more difficult ones, such as recognition of a person from images acquired years apart.

CHAPTER 3

Human Recognition Using Gait

Human gait is a spatio-temporal phenomenon that characterizes the motion characteristics of an individual. Study of human gait and it mathematical modeling has implications in a large number of disciplines from surveillance to medicine to the entertainment industry. When person identification is attempted in natural settings such as those arising in surveillance applications, biometrics such as fingerprint or iris are no longer applicable. Furthermore, night vision capability (an important component in surveillance) is usually not possible with these biometrics. Even though an IR camera would reveal the presence of people, the facial features are far from discernible in an IR image at large distances. The attractiveness of gait as a biometric arises from the fact that it is nonintrusive and can be detected and measured even in low-resolution video. Furthermore, it is harder to disguise than static appearance features such as face and it does not require a cooperating subject. In kinesiology, models for human walk can be used for early diagnosis of different disorders and can be used to supplement information available from other sources like X rays. They can be used to monitor the recovery of individuals after surgery or injuries. Human motion modeling has implications for the entertainment and communication industries also.

Study of human gait is a relatively new area for computer vision researchers. However, extensive work has been carried out in the psychophysics community on the ability of humans to recognize others by their style of walking. We will review some of their work in the next section. In the computer vision community, research on gait has concentrated mostly on recognition algorithms. These methods can be divided into two groups: appearance based and model based. Alternatively, these methods can be classified as deterministic or stochastic. Deterministic appearance based methods are discussed in (31; 80; 128), while well-known stochastic methods in the same category use hidden markov models (HMM) (95; 162). Model-based methods are fewer largely because of the difficult of obtaining accurate 3D models of the human body. In a later chapter, we will discuss methods for modeling human gait using kinematic chains and multiple cameras.

3.1   REVIEW OF EXISTING WORK

Early research on gait can be primarily grouped into two categories. The first category involves psychophysical studies of gait, viz. studying the ability of human observers to recognize gait. The second category includes biomechanical studies, viz. studying kinetics and kinematics associated with gait. These studies in part led to the development of computational approaches to gait-based human identification. In this section we describe some of the gait research done in the areas of psychophysics, biomechanics, and computer vision.

3.1.1   Studies in Psychophysics

The belief that humans can distinguish between gait patterns of different individuals is widely held. Intuitively, it is possible to think of the qualities of walk that help a perceiver identify an approaching figure even before the face becomes discernible. These gait-related quantities include stride length, bounce, rhythm, speed, and perhaps even attributes such as swagger or body swing. The suggestion that humans can identify people by their gait was investigated in a series of early studies by Johansson (77). Kozlowski and Cutting (88) first investigated whether perceivers could identify the gender of a walker from point light displays attached to the body. Their results indicated an accuracy rate of 65% and 70% when the walker was viewed from the side. A minimum exposure time of 2 second was required for gender identification. In (36), it was suggested that gender may be identified indirectly through a determination of the "center of moment" of a walker. The center of moment is the point about which all movement in all body parts has regular geometric relations and as a result of the differences in the shoulder and hip widths in men and women, the center of moment is higher in females than males.

The demonstration that gender could be extracted from gait provided insight into how perceivers might discriminate between gait patterns of different individuals. The prospect for perceivers then being able to identify individuals from their gaits was thus encouraging. Cutting and Kozlowski (35) demonstrated that perceivers could reliably recognize themselves and their friends from dynamic point-light displays. Barclay *et al.* (6) suggested that individual walking styles might be captured by differences in a basic series of pendular limb motions. Interestingly, Beardsworth and Buckner (9) have shown that the ability to recognize oneself from a point-light display is greater than the ability to recognize one's friends, despite the fact that we rarely see our own gait from a third-person perspective. This suggests that, as well as the apparent sensitivity of the human visual system to biological motion, it is likely that there is some transfer of information from the kinematic modality to the visual modality.

Stevenage *et al.* (160) also explored the ability of people to identify people using gait information alone. In one experiment, people were given video footage of six walking subjects to study. After the studying phase was over, one group viewed the same walkers under simulated daylight in which the silhouette and motion of the walkers was clearly visible. A second group viewed the subjects under simulated dusk in which the outline and motion were difficult to see. A third group was shown only the point light displays of the walkers. It was found that all the three groups were able to learn the gait of the six walkers and label their identity regardless of the nature of data. In a second experiment, a different set of viewers were given only 2 s to view a target, and they had to pick the target from a parade of walkers under similar viewing conditions as the first experiment. It was found that even with such a brief exposure time and their unfamiliarity with the walking subjects, the observers could identify the target correctly at greater than chance rate.

3.1.2   Medical Studies

Although walking is one of the most universal of all human activities, it was only in the early part of the twentieth century that the detailed components of this act started being systematically examined. The studies from the Greek era until the mid-nineteenth century were, for the most part, observational. The 1950s saw many diversified kinematic gait studies, most prominently at the University of California. These studies were spurred by a need for an improved understanding of locomotion for the treatment of World War II veterans. One of the goals of gait researchers was to build a model for normal gait, deviations from which could be used to study gait abnormalities. Such studies involved collecting the gait of a large number of subjects with no gait abnormalities. The presence of identity information in gait was a by-product of these studies. In (115), subjects with reflective targets attached to specific anatomical landmarks walk before a camera in the illumination of a strobe light flashing 20 times per second. A mirror was mounted over the walking area so that the target images projected in the overhead view are captured as well. Twenty different components of gait, which include motions of the hip, knee, ankle, etc., were studied and it was found that when considered together, the gait of the subjects was unique. Certain components such as pelvic and thoracic rotation had more pronounced interperson difference. The ability of recognizing and or modifying specific movement patterns is of particular interest for successful and effective therapeutic interventions. A recent study by Schollhorn *et al.* (147) studied the gait of fifteen subjects to study the presence of identity information in gait. It was found in this study that kinetic variables (captured using a force platform) as well as kinematic variables (captured by reflective markers on the thigh, shank, and hip) were both necessary for gait identification. Furthermore simply using the leg portion of the body was adequate for getting good identification performance.

3.1.3   Computational Approaches

Medical and psychophysical studies, as discussed in Sections 3.1.1 and 3.1.2 indicate that there exists identity information in gait. It is therefore interesting to study the utility of gait as a biometric. In recent years, there has been an increase in research related to gait-based human recognition. We give a summary of some of the examples below, but the listing is by no means complete.

As noted before, joint angles may be sufficient for recognizing people by their gait. However reliably recovering joint angles from a monocular video is a hard problem. Furthermore, using only joint angle information ignores structural traits associated with an individual such as height, girth, etc. It is therefore, reasonable to include appearance as a part of the gait recognition features. Approaches to gait recognition problem can be broadly classified as being either model based or model free. Both methodologies follow the general framework of feature extraction, feature correspondence and high-level processing. The major difference is with regard to feature correspondence between two consecutive frames. Methods that assume *a priori* models match the 2D image sequences to the model data. Feature correspondence is automatically achieved once matching between the images and the model data is established. Examples of this approach include the work of Lee *et al.* (96), where several ellipses are fitted to different parts of the binarized silhouette of the person and the parameters of these ellipses such as location of its centroid, eccentricity, etc., are used as a feature to represent the gait of a person. Recognition is achieved by template matching. In (34), Cunado *et al.* extract a gait signature by fitting the movement of the thighs to an articulated pendulum-like motion model. The idea is somewhat similar to an early work by Murray (115) who modeled the hip rotation angle as a simple pendulum, the motion of which was approximately described by simple harmonic motion. Model-free methods establish correspondence between successive frames based upon the prediction or estimation of features related to position, velocity, shape, texture, and color. Alternatively, they assume some implicit notion of what is being observed. Examples of this approach include the work of Huang *et al.* (70), where optical flow is first derived from a motion sequence for the duration of a walk cycle. Principal components analysis is then applied to the binarized silhouette to derive what are called eigen gaits. Benabdelkader *et al.* (26) use image self-similarity plots as a gait feature. Little and Boyd (102) extract frequency and phase features from moments of the motion image derived from optical flow and use template matching to recognize different people by their gait. A dynamic time warping (DTW) (51) based algorithm for gait recognition was proposed in (80). The algorithm matches two gait sequences (probe and gallery) by computing the distance, as a function of time, between two feature sets representing the data. This approach can be used even when substantial training data is not available. It can also account for modest variation in speed of walking. Two of the most successful approaches, till date, in gait recognition are reported in (162) and (169). In (162), the authors used a HMM (134) to represent the gait of each individual. This algorithm will be described in detail in the next section. A method for identifying individuals by shape, which is automatically extracted from a cluster of similar poses obtained from a spectral partitioning framework, was proposed in (169). Most of the above methods rely on the availability of a side view in order to extract the gait parameters. Two approaches for gait-based human recognition when side views are not available are presented in (61) and (79). A study of the role of shape and kinematics in computer vision based gait recognition approaches was recently presented in (178).

3.1.4   Evaluation of Gait Recognition Algorithms

Similar to the FERET evaluations for face recognition, a HumanID Gait Challenge Problem was introduced in order to measure progress of different recognition algorithms (129) (http://www.gaitchallenge.org). The challenge problem consists of a baseline algorithm, a set of 12 experiments and a data set of 122 people. The baseline algorithm estimates silhouettes by background subtraction, and performs recognition by temporal correlation of silhouettes. Twelve experiments examine the effects of five covariates: change of viewing angle, change in shoe type, change in

TABLE 3.1: Probe sets for the gait challenge data

EXPERIMENT	PROBE DESCRIPTION (SURFACE C/G, SHOE A/B, CAMERA L/R, CARRY NB/BF, TIME)	NUMBER OF SUBJECTS
A	(G, A, L, NB, T1+T2)	122
B	(G, B, R, NB, T1+T2)	54
C	(G, B, L, NB, T1+T2)	54
D	(C, A, R, NB, T1+T2)	121
E	(C, B, R, NB, T1+T2)	60
F	(C, A, L, NB, T1+T2)	121
G	(C, B, L, NB, T1+T2)	60
H	(G, A, R, BF, T1+T2)	120
I	(G, B, R, BF, T1+T2)	60
J	(G, A, L, BF, T1+T2)	120
K	(G, A/B, R, NB, T2)	33
L	(G, A/B, R, NB, T2)	33

Note: The gallery is (G, A, R, NB, T1+T2). C/G represents concrete/grass surface, L/R represents left or right camera, NB/BF represents carrying a briefcase or not, T1 and T2 represent the data collected at two different time instants.

walking surface, carrying or not carrying a briefcase, and temporal differences. A description of the different experiments (probe sets) is given in Table 3.1:. Identification and verification scores for all the experiments are reported using the baseline algorithm. The relative performance of the HMM based method with the baseline will be presented in the next section.

3.2   HMM FRAMEWORK FOR GAIT RECOGNITION

The HMM framework is suitable because the gait of an individual can be visualized as his/her adopting postures from a set, in a sequence that has an underlying structured probabilistic nature. The postures that the individual adopts can be regarded as the states of the HMM and are typical to that individual and provide a means of discrimination. The framework assumes that, during a walk cycle, the individual transitions among N discrete postures or states. An adaptive filter is used to automatically detect the cycle boundaries. The method is not dependent on the particular feature vector used to represent the gait information contained in the postures. The statistical nature of the HMM lends robustness to the model. In the method described below, the binarized background-subtracted image is used as the feature vector and different distance metrics, such as those based on the L_1 and L_2 norms of the vector difference, and the normalized inner product of the vectors, are used to measure the similarity between feature vectors.

3.2.1   Overview of the HMM Framework

Let the database consists of video sequences of P persons. The model for the pth person is given by $\lambda_p = (A_p, B_p, \pi_p)$ with N number of states. The model, λ_p, is built from the observation sequence for the pth person using the sequence of feature vectors given by $\mathcal{O}_p = \{\mathbf{O}_1^p, \mathbf{O}_2^p, \ldots, \mathbf{O}_{T_p}^p\}$, where T_p is the number of frames in the sequence of the pth person. A_p is the transition matrix and π_p is the initial distribution. The B_p parameter consists of the probability distributions for a feature vector conditioned on the state index, i.e., the set $\{P_1^p(.), P_2^p(.), \ldots, P_N^p(.)\}$. The probability distributions are defined in terms of *exemplars*, where the jth exemplar is a typical realization of the jth state. The exemplars for the pth person are given by $\mathcal{E}_p = \{\mathbf{E}_1^p, \mathbf{E}_2^p, \ldots, \mathbf{E}_N^p\}$. Henceforth, the superscript denoting the index of the person is dropped for simplicity. The motivation behind using an exemplar-based model is that the recognition can be based on the distance measure between the observed feature vector and the exemplars. The distance metric is evidently a key factor in the performance of the algorithm. $P_j(\mathbf{O}_t)$ is defined

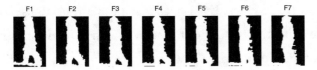

FIGURE 3.1: Part of an observation sequence.

as a function of $D(\mathbf{O}_t, \mathbf{E}_j)$, the distance of the feature vector $\mathbf{O}_t$ from the jth exemplar.

$$P_j(\mathbf{O}_t) = \alpha e^{-\alpha D(\mathbf{O}_t, \mathbf{E}_j)} \qquad (3.1)$$

During the *training* phase, a model is built for all the subjects, indexed by $p = 1, 2, \ldots, P$, in the gallery. An initial estimate of $\mathcal{E}_p$ and λ_p is formed from $\mathcal{O}_p$, and these estimates are refined iteratively. Note that B is completely defined by $\mathcal{E}$ if α is fixed beforehand. We can iteratively estimate A and π by using the Baum–Welch algorithm (41), keeping $\mathcal{E}$ fixed. The algorithm to reestimate $\mathcal{E}$ is determined by the choice of the distance metric. During *testing*, given a gallery $\mathcal{L} = \{\lambda_1, \lambda_2, \ldots, \lambda_P\}$ and the probe sequence of length T, $\mathcal{X} = \{\mathbf{X}_1, \mathbf{X}_2, \ldots, \mathbf{X}_T\}$ traversing the path $\mathcal{Q} = \{q_1, q_2, \ldots, q_T\}$, q_t being the state index at time t, we obtain the ID of the probe sequence as

$$\text{ID} = \arg_p \max_{\mathcal{Q}, p} \Pr[\mathcal{Q} \,|\, \mathcal{X}, \lambda_p]. \qquad (3.2)$$

The feature vector used is the binarized version of the background subtracted images. The images are scaled and aligned to the center of the frame as in Fig. 3.1, which features part of a sequence of feature vectors. We now describe the methods used to obtain initial estimates of the HMM parameters, the training algorithm and finally, identification results using USF data described in (129).

3.2.2 Initial Estimate of HMM Parameters

In order to obtain a good estimate of the exemplars and the transition matrix, we first obtain an initial estimate of an ordered set of exemplars from the sequence and the transition matrix and successively refine the estimate. The initial estimate for the exemplars, $\mathcal{E}^0 = \{\mathbf{E}_1^0, \mathbf{E}_2^0, \ldots, \mathbf{E}_N^0\}$ is such that the only transitions allowed are from the jth state to either the jth or the $(j \bmod N + 1)$th state. A corresponding initial estimate of the transition matrix, A^0 (with $A_{j,j}^0 = A_{j,j \bmod N+1}^0 = 0.5$, and all other $A_{j,k}^0 = 0$) is also obtained. The initial probabilities π_j are set to be equal to $1/N$.

We observe that the gait sequence is quasi-periodic and we use this fact to obtain the initial estimate $\mathcal{E}^0$. We divide the sequence into "cycles," where a cycle is defined as that segment of the sequence bounded by silhouettes where the subject has arms by his/her side and legs approximately aligned with each other. We can further divide each cycle into N temporally adjacent clusters of approximately equal size. We visualize the frames of the jth cluster of all cycles to be generated from the jth state. Thus we get a good initial estimate of $\mathbf{E}_j$ from the feature vectors belonging to the jth cluster. For example, assume that the training sequence is given by $\mathcal{Y} = \{\mathbf{Y}_1, \mathbf{Y}_2, \ldots, \mathbf{Y}_T\}$. We partition the sequence into K cycles, with the kth cycle given by frames in the set $\mathcal{Y}_k = \{\mathbf{Y}_{S_k}, \mathbf{Y}_{S_k+1}, \ldots, \mathbf{Y}_{S_k+L_k-1}\}$, where S_k and L_k are the index of the first frame of the kth cycle and the length of the kth cycle, respectively. We define the first cluster to comprise of frames with indices $S_k, S_k + 1, \ldots, S_k + \frac{1}{2}L_k/N, S_k + L_k - \frac{1}{2}L_k/N, S_k + L_k - \frac{1}{2}L_k/N + 1, \ldots, S_k + L_k - 1$. The jth cluster ($j = 2, 3, \ldots, N$) is defined to comprise of frames with indices $S_k + (j - \frac{3}{2})L_k/N, S_k + (j - \frac{3}{2})L_k/N + 1, \ldots, S_k + (j - \frac{1}{2})L_k/N$. We need to robustly estimate the cycle boundaries so that we can partition the sequence into N clusters and obtain the initial estimates of the exemplars. If the sums of the foreground pixels of each image are plotted with respect to time, then, as per our definition of a cycle, the minimum values should correspond to the cycle boundaries. We denote the sum of the foreground pixels of the silhouette in the nth frame as $s[n]$. This signal is noisy and may contain several spurious minima. However we can exploit the quasi-periodicity of the signal and filter the signal to remove the noise before identifying the minimum values. Methods such as median filtering or differential smoothing of $s[n]$ are not very robust as they do not take into account the frequency of the gait.

The specifications of the bandpass filter are such as to allow frequencies that are typical for a fast walk. The video is captured at 30 frames per second, and the sampling frequency, $f_s = 1/30$ and $T_s = 30$. The maximum gait frequency is assumed to be $f_m = 0.1$ corresponding to a cycle period of $T_m = 10$. A Hamming window of length L is used. The extended sequence $x[n]$ is obtained by symmetrically extending $s[n]$ in both directions by $L/2$. Therefore the sequence $x[n]$ has length $M = N + L$. The resultant sequence is filtered using a bandpass filter (with upper cut-off frequency $f_{uc} = f_m$), in both directions to remove phase delay. The distances between the locations where minimum values occur of the filtered sequence lead us to an estimate of the cycle period. The cycle frequency is estimated as the inverse of the median of cycle periods. Using this revised estimate of the frequency of the gait, $\hat{f}$, a new filter is constructed with upper cut off frequency $f_{uc} = \hat{f} + 0.02$. A manual examination of all the sequences in the gallery in the gait challenge database revealed a 100% detection rate with hardly any false detection of cycle boundaries.

3.2.3   Training the HMM Parameters

The iterative refinement of the estimates is performed in two steps. In the first step, a Viterbi evaluation (134) of the sequence is performed using the current values for the exemplars and the transition matrix. Thus feature vectors are clustered according to the most likely state they originated from. The exemplars for the states are newly estimated from these clusters. Using the current values of the exemplars, $\mathcal{E}^{(i)}$ and the transition matrix, $A^{(i)}$, Viterbi decoding is performed on the sequence $\mathcal{Y}$ to obtain the most probable path $\mathcal{Q} = \{q_1^{(i)}, q_2^{(i)}, \ldots, q_T^{(i)}\}$, where $q_t^{(i)}$ is the state at time t. Thus the set of observation indices, whose corresponding observation is estimated to have been generated from state j is given by $T_j^{(i)} = \{t : q_t^{(i)} = j\}$. We now have a set of frames for each state and we would like to select the exemplars so as to maximize the probability in (3.3). If we use the definition in (3.1), (3.4) follows:

$$\mathbf{E}_j^{(i+1)} = \arg_{\mathbf{E}} \max \prod_{t \in T_j^{(i)}} P(\mathbf{Y}_t \mid \mathbf{E}) \qquad (3.3)$$

$$\mathbf{E}_j^{(i+1)} = \arg_{\mathbf{E}} \min \sum_{t \in T_j^{(i)}} D(\mathbf{Y}_t, \mathbf{E}). \qquad (3.4)$$

The actual method for minimizing the distance in (3.4) however depends on the distance metric used. We have experimented with three different distance measures, namely the Euclidean (EUCLID) distance, the inner product (IP) distance, and the sum of absolute difference (SAD) distance, which are given by (3.5)–(3.7), respectively. Note that though $\mathbf{Y}_t$ and $\mathbf{E}$ are 2D images, they are represented as vectors of dimension $D \times 1$ for ease of notation. $\mathbf{1}_{D \times 1}$ is a vector of D ones:

$$D_{s\,\text{EUCLID}}(\mathbf{Y}, \mathbf{E}) = (\mathbf{Y} - \mathbf{E})^T(\mathbf{Y} - \mathbf{E}) \tag{3.5}$$

$$D_{s\,\text{IP}}(\mathbf{Y}, \mathbf{E}) = 1 - \frac{\mathbf{Y}^T\mathbf{E}}{\sqrt{\mathbf{Y}^T\mathbf{Y}\mathbf{E}^T\mathbf{E}}} \tag{3.6}$$

$$D_{s\,\text{SAD}}(\mathbf{Y}, \mathbf{E}) = |\mathbf{Y} - \mathbf{E}|^T \mathbf{1}_{D \times 1}. \tag{3.7}$$

The equations for updating the jth element of the exemplars in the EUCLID distance, IP distance, and the SAD distance cases are presented in (3.8)–(3.10), respectively. $\tilde{\mathbf{Y}}$ denotes the normalized vector $\mathbf{Y}$ and $|\mathcal{T}_j^{(i)}|$ denotes the cardinality of the set $\mathcal{T}_j^{(i)}$:

$$\mathbf{E}_j^{(i+1)}(j) = \frac{1}{|\mathcal{T}_j^{(i)}|} \sum_{t \in \mathcal{T}_j^{(i)}} \mathbf{Y}_t(j) \tag{3.8}$$

$$\mathbf{E}_j^{(i+1)}(j) = \sum_{t \in \mathcal{T}_j^{(i)}} \tilde{\mathbf{Y}}_t(j) \tag{3.9}$$

$$\mathbf{E}_j^{(i+1)}(j) = \text{median}_{t \in \mathcal{T}_j^{(i)}} \{\mathbf{Y}_t(j)\}. \tag{3.10}$$

The exemplars estimated for one observation sequence using the three distance metrics in (3.5)–(3.7) are displayed in Fig. 3.2. Given $\mathcal{E}^{(i+1)}$ and $A^{(i)}$, we can calculate $A^{(i+1)}$ using the Baum–Welch algorithm (134). Thus we can successively refine our estimates of the HMM parameters. It usually takes only a few iterations in order to obtain an acceptable estimate.

3.2.4 Identifying from a Test Sequence

Identifying a sequence involves deciding which of the model parameters to use for discrimination parameters. Given the models in the gallery, $\mathcal{L} = \{\lambda_1, \lambda_2, \ldots, \lambda_P\}$, and the probe sequence, $\mathcal{X} = \{\mathbf{X}_1, \mathbf{X}_2, \ldots, \mathbf{X}_T\}$, we would like to find the model and the path that maximizes the probability of the path given the probe sequence. The ID is obtained as in (3.2).

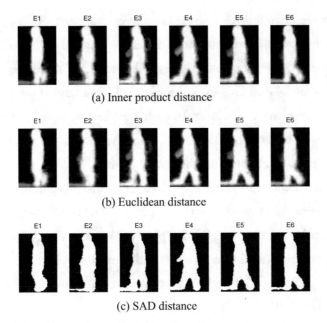

(a) Inner product distance

(b) Euclidean distance

(c) SAD distance

FIGURE 3.2: Exemplars estimated using various distance measures.

We do not need to use the trained parameter set, λ, as a whole. For example, if we believe that the transition matrix is predominantly indicative of the speed at which the subject walks, and is therefore not suitable as a discriminant of the ID of the subject, then we have the option of using only part of the parameter set given by $\gamma_p = (B_p, \pi_p)$ instead of using the HMM parameter set in its entirety. In this case, the conditional probability of the sequence, given the ID, is given as follows. The Baum–Welch algorithm could be used in order to obtain $A_p^{\mathcal{X}}$ recursively in (3.12):

$$\Pr[\mathcal{Q} \mid \mathcal{X}, \gamma_p] = \Pr[\mathcal{Q} \mid \mathcal{X}, A_p^{\mathcal{X}}, \gamma_p] \tag{3.11}$$

$$A_p^{\mathcal{X}} = \arg_A \max \Pr[\mathcal{X} \mid A, \gamma_p]. \tag{3.12}$$

3.2.5 Experimental Results

The objective of our experiments was to evaluate the performance of the algorithm and also compare the efficacy of the different distance measures in gauging the similarity between two images as far as posture is concerned. As described before, the gait challenge or USF database contains video sequences of 122 individuals, different sets of them appearing in sequences collected under 12 different conditions. The sequences are labeled Gallery and Probe A–L. We trained our parameters on the sequences from the Gallery set. In each experiment, we tried to identify the sequences in each of the seven Probe sets from the parameters obtained from the Gallery set using the inner product distance measure. The ID was calculated using (3.2). The experiments were repeated with different distance measures. The results of the experiment using the IP distance measure between feature vectors in the form of cumulative match scores (CMS) plots (128) are in Fig. 3.3(a). We observe that the distance measure that works best and is most simple to implement is the inner product distance. The performance comparison with the baseline (128) is illustrated in Fig. 3.3(b).

From the experiments we note that the biggest drop in performance occurs due to change in surface type and when there is a difference in time between the gallery and the probe. Reasons for the sudden drop are not yet fully understood. Probable causes may be change in the silhouette (especially lower part for surface change) and change in clothing due to time differences. Note that the performance does not change much with small changes in viewing direction. In summary, gait recognition under arbitrary conditions is still an open research problem.

3.3 VIEW INVARIANT GAIT RECOGNITION

The gait of a person is best reflected when he or she presents a side view (referred to in this chapter as a canonical view) to the camera. Hence, most of the above gait recognition algorithms rely on the availability of the side view of the subject. The situation is analogous to face recognition where it is desirable to have frontal views of the person's face. In realistic scenarios, however, gait recognition algorithms need to work in a situation where the person walks at an arbitrary angle to the camera. The most general solution to this problem is to estimate the 3D model for the person. Features extracted from the 3D model can then be used to provide the gait model for the person. This problem requires the solution of the SfM or

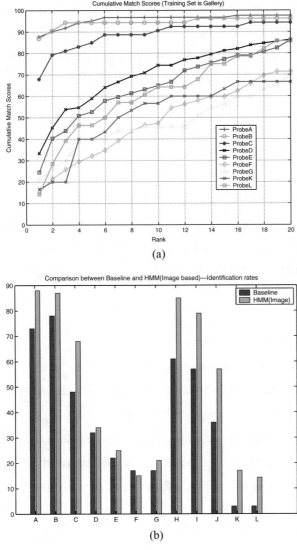

FIGURE 3.3: (a): CMS plots of Probes A–L tested against Gallery. (b) Comparison of identification rates of HMM and baseline algorithm.

stereo reconstruction problems (45; 65), which are known to be hard for articulating objects. In the absence of methods for recovery of accurate 3D models, a simple way to exploit existing appearance-based methods is to synthesize the canonical views of a walking person. In (61), Shakhnarovich *et al.* compute an image based visual hull from a set of monocular views, which is then used to render virtual canonical views for tracking and recognition. Gait recognition is achieved by matching a set of image features based on moments extracted from the silhouettes of the synthesized probe video to the gallery. An alternative to synthesizing canonical views is the work of Bobick and Johnson (17). In this work, two sets of activity-specific static and stride parameters are extracted for different individuals. The expected confusion matrix for each set is computed to guide the choice of parameters under different imaging conditions (viz. indoor vs. outdoor, side-view vs. angular-view, etc.). A cross-view mapping function is used to account for changes in viewing direction. The set of stride parameters (which is smaller than the set of static parameters) is found to exhibit greater resilience to viewing direction. A method for recognizing the gait of an individual using joint angle trajectories was presented in (167). However, representation using such a small set of parameters may not give good recognition rates on large databases.

3.3.1   Overview of Algorithm

We present a view-invariant gait recognition algorithm for the single camera case (79; 81). We show that it is possible to synthesize a canonical view from an arbitrary one without explicitly computing the 3D depth. Consider a person walking along a straight line that subtends an angle θ with the image plane (AC in Fig. 3.5). If the distance, Z_0, of the person from the camera is much larger than the width, ΔZ, of the person, then it is reasonable to replace the scaling factor $\frac{f}{Z_0 + \Delta Z}$ for perspective projection by an average scaling factor $\frac{f}{Z_0}$. In other words, for human identification at a distance, we can approximate the actual 3D human as a planar object. Assume that we are given a video of a person walking at a fixed angle θ (Fig. 3.5). We show that by tracking the direction of motion, α, in the video sequence, we can estimate the 3D angle θ. This can be done by using the optical flow based SfM equations. Under the assumption of planarity, knowing angle θ and the calibration parameters, we can synthesize side views of the sequence of images of an unknown walking person without explicitly computing the 3D model of the person. We refer to this approach as the "implicit SfM" approach. In the case where there is no

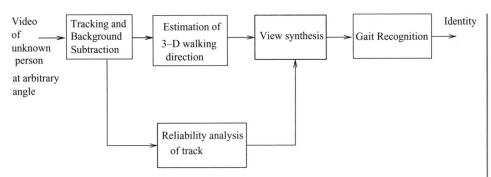

FIGURE 3.4: Framework for view invariant gait recognition.

real translation of the person, e.g. person walking on a treadmill, an alternative approach is employed to obtain the synthesized views of the person. Given a set of point correspondences for a planar surface between the canonical and noncanonical views in a set of training images, we compute a homography. This homography is then applied to the binary silhouette of the person to obtain the synthesized views. We refer to this approach as the "homography approach."

An overview of our gait recognition framework is given in Fig. 3.4. We present recognition performance using two publicly available gait databases (NIST and CMU). The implicit SfM approach is used for the NIST databases, while the homography approach is used for the CMU database. Keeping in view the limited quantity of training data, the DTW algorithm (80) is used for gait recognition. A by-product of the above method is a simple algorithm to synthesize novel views of a planar scene.

3.3.2 Theory for View Synthesis

In the imaging setup shown in Fig. 3.5, the coordinate frame is attached rigidly to a camera with the origin at the center of perspective projection and the Z-axis perpendicular to the image plane. Assume that the person walks with a translational velocity $\mathbf{V} = [v_X, 0, v_Z]^T$ along the line AC. The line AB is parallel to the image plane XY and this is the direction of the canonical view that needs to be synthesized. The angle between the straight line AB and AC, i.e. θ, represents a rotation about

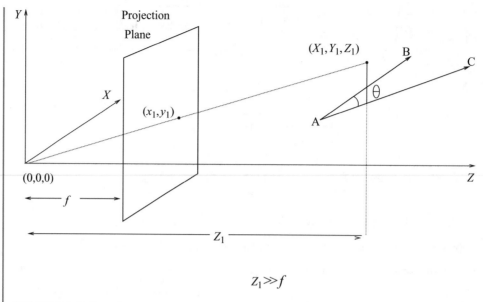

FIGURE 3.5: Imaging geometry.

the vertical axis referred to as the azimuth angle. We will use the notation that $[X, Y, Z]$ denotes the coordinates of a point in 3D and $[x, y]$ its projection on the image plane.

3.3.2.1 Tracking

We assume that only one subject is present in the field of view and the availability of a high-level motion detection module, which identifies abrupt changes in direction of motion and segments of *approximately* constant heading directions. Assuming that we can find the location $(x_{\text{ref}}, y_{\text{ref}})$ of the persons head at the start of such a segment, we use a sequential Monte Carlo particle filter (73) to track the head of the person to get $\{(x^i(t), y^i(t)), w^i(t)\}$ where the superscript denotes the index of the particle and $w^i(t)$ denotes the probability weight for the estimate $((x^i(t), y^i(t))$.

3.3.2.2 Estimation of 3D Azimuth Angle

Assume that the motion between two consecutive frames in the video sequence is small. Using the optical flow based SfM equations, let $p(x(t), y(t))$ and $q(x(t), y(t))$

represent the horizontal and vertical velocity fields of a point $(x(t), y(t))$ (e.g. centroid of the head) in the image plane, where t denotes time. Since we consider a straight line motion along AC, p and q are related to the 3D object motion and scene depth by (118):

$$p(x(t), y(t)) = (x(t) - f x_f(t)) h(x(t), y(t)) \tag{3.13}$$

$$q(x(t), y(t)) = y(t) h(x(t), y(t)), \tag{3.14}$$

where f denotes the focal length, $h(x(t), y(t)) = v_Z(t)/Z(x(t), y(t))$ is the scaled inverse scene depth and $x_f(t) = \cot(\theta(t)) = \frac{v_X}{v_Z}(t)$, $y_f(t) = \frac{v_Y}{v_Z}(t)$ is the focus of expansion (FOE). When $v_Z = 0$ but $v_X \neq 0$, we see that $\theta = 0$, i.e. the canonical direction of walk, AB. Also, in this case $q(x, y) = 0$.

For the constant velocity models, $v_Z(t) = v_Z(\neq 0)$ and $v_X(t) = v_X(\neq 0)$, $\cot(\theta(t)) = v_X/v_Z$. In this case, given the initial position of the tracked point $(x_{\text{ref}}, y_{\text{ref}})$, it can be shown that (79)

$$\cot(\alpha)(x_{\text{ref}}, y_{\text{ref}}) = \frac{p(x(t), y(t))}{q(x(t), y(t))} = \frac{x_{\text{ref}} - f \cot(\theta)}{y_{\text{ref}}}. \tag{3.15}$$

Thus, given f and (x_0, y_0), the azimuth angle θ can be computed as

$$\cot(\theta) = \frac{x_{\text{ref}} - y_{\text{ref}} \cot(\alpha(x_{\text{ref}}, y_{\text{ref}}))}{f}. \tag{3.16}$$

Knowing (x_0, y_0), $\cot(\alpha)$, and θ, f can be computed as part of a calibration procedure.

3.3.2.3 Statistical Error Analysis of Azimuth Estimation

Let $f_X(x)$ denote the distribution of a random variable X. Defining $r = \cot(\alpha)$, from (3.15) we have $r = p/q$. In order to obtain the distribution of r we define an auxiliary random variable $s = q$. Then it can be shown using the properties of a bijective transformation of a pair of random variables (122) that $f_{RS}(r, s) = f_{PQ}(rs, s)|s|$. The probability distribution function (pdf) of r follows by marginalization as

$$f_R(r) = \int_{-\infty}^{\infty} f_{RS}(r, s)\, ds = \int_{-\infty}^{\infty} |s| f_{PQ}(rs, s)\, ds. \tag{3.17}$$

In general, computing the above integral is nontrivial. We derive expressions for the pdf of r for the following special cases:

1. Uniform additive noise for p and q: Given $p = \bar{p} + n_p$ and $q = \bar{q} + n_q$, where $n_p \sim U(-\Delta_1, \Delta_1)$, $n_q \sim U(-\Delta_2, \Delta_2)$ and $\bar{p}$ and $\bar{q}$ denote the true image-plane velocities, it can be shown that

$$f_R(r) = \frac{1}{4\Delta_1\Delta_2} \int_{I_s} |s|\, ds, \tag{3.18}$$

   where $I_s = \{v : \bar{p} - \Delta_1 \leq s \leq \bar{p} + \Delta_1 \bigcap (\bar{q} - \Delta_2)r \leq s \leq (\bar{q} + \Delta_2)r\}$.

2. Zero mean multiplicative Gaussian noise for p and q: Given $p = \bar{p}n_p$ and $q = \bar{q}n_q$, where $n_p \sim \mathcal{N}(0, \sigma_1^2$ and $n_q \sim \mathcal{N}(0, \sigma_2^2)$ ($\bar{p}$ and $\bar{q}$ denote the true image-plane velocities), we can explicitly solve (3.17) to get $f_R(r) = \frac{\sigma_1\sigma_2}{\pi} \frac{1}{\sigma_2^2 r^2 + \sigma_1^2}$ viz. Cauchy with scale parameter $\frac{\sigma_1}{\sigma_2}$.

From the standpoint of implementation, under arbitrary noise distributions, the sequential Monte Carlo filter can be used to get the distribution of $\cot(\alpha)$. In order to estimate the distribution of $\cot(\alpha)$, we consider time-instants separated by the approximate period (T) of the walk cycle. Thus

$$\cot(\alpha(t + T)) \sim \left(\frac{x^i(t + T) - \tilde{x}(t)}{y^i(t + T) - \tilde{y}(t)}, w^i(t + T) \right), \tag{3.19}$$

where $(\tilde{x}(t), \tilde{y}(t)) = \arg\max_w^i (x^i(t), y^i(t), w^i(t))$. The distribution for $\cot(\theta)$ follows from (3.16) using the theory of propagation of random variables (122).

3.3.2.4 View Synthesis

Having obtained the angle θ, we synthesize the canonical view. Let Z denote the distance of the object from the image plane. If the dimensions of the object are small compared to Z, then the variation in θ, $d\theta \approx 0$. This essentially corresponds to assuming a planar approximation to the object. Let $[X_\theta, Y_\theta, Z_\theta]'$ denote the coordinates of any point on the person (as shown in the Fig. 3.5) who is walking at an angle $\theta \geq 0$ to the plane passing through the starting point $[X_{\text{ref}} Y_{\text{ref}} Z_{\text{ref}}]'$ and parallel to the image plane, which we shall refer to, hereafter, as the canonical plane. Computing the 3D coordinates of the synthesized point involve a rotation about the line passing through the starting point.

Then

$$
\begin{bmatrix} X_0 \\ Y_0 \\ Z_0 \end{bmatrix} = R(\theta) \begin{bmatrix} X_\theta - X_{\text{ref}} \\ Y_\theta - Y_{\text{ref}} \\ Z_\theta - Z_{\text{ref}} \end{bmatrix} + \begin{bmatrix} X_{\text{ref}} \\ Y_{\text{ref}} \\ Z_{\text{ref}} \end{bmatrix}, \tag{3.20}
$$

where

$$
R(\theta) = \begin{bmatrix} \cos(\theta) & 0 & \sin(\theta) \\ 0 & 1 & 0 \\ -\sin(\theta) & 0 & \cos(\theta) \end{bmatrix}. \tag{3.21}
$$

Denoting the corresponding image plane coordinates as $[x_\theta, y_\theta]'$ and $[x_0, y_0]'$ (for $\theta = 0$) and using the perspective transformation, we can obtain the equations for $[x_0, y_0]'$ as (79)

$$
\begin{aligned}
x_0 &= f \frac{x_\theta \cos(\theta) + x_{\text{ref}}(1 - \cos(\theta))}{-\sin(\theta)(x_\theta + x_{\text{ref}}) + f} \\
y_0 &= f \frac{y_\theta}{-\sin(\theta)(x_\theta + x_{\text{ref}}) + f},
\end{aligned} \tag{3.22}
$$

where $x = f(X/z)$ and $y = f(Y/z)$. Equation (3.22) is attractive since it does not involve the 3D depth; rather it is a direct transformation of the 2D image plane coordinates in the noncanonical view to get the image plane coordinates in the canonical one. Thus using the estimated azimuth angle θ we can obtain a synthetic canonical view using (3.22). In summary, obtaining the synthesized views from the noncanonical views involves a three-step procedure:

1. Estimation of the image plane angle α from the video.

2. Estimation of the azimuth angle θ using (3.16).

3. View synthesis using (3.22).

3.3.2.5 Alternative Approach for View Synthesis

The above procedure relies on the true translation of the walking person in the video. For the case when there is no translation (e.g. person walking on a treadmill), neither α nor f can be estimated from the video. An alternative approach to view synthesis can be used in this case. We observe that (3.22) is a homography of the form

$$H(\theta) = \begin{bmatrix} a & 0 & b \\ 0 & c & 0 \\ d & 0 & e \end{bmatrix}, \tag{3.23}$$

where $a = f\cos(\theta)$, $b = fx_{\text{ref}}(1 - \cos(\theta))$, $c = f$, $d = -\sin(\theta)$, and $e = -x_{\text{ref}}\sin(\theta) + f$. Hence if we have point correspondences for any planar surface between the canonical and noncanonical views in a set of training images, we can estimate a homography of the type in (3.23). This homography can then be applied to the binary silhouette of the person in the noncanonical view to get the synthesized views.

3.3.3   Experimental Results

A DTW algorithm is used to match two gait sequences, one from the gallery and the other from the probe. As required, either one or both may be synthesized depending upon the viewing direction of the camera.

The NIST database consists of 30 people walking along a Σ-shaped walking pattern. There are two cameras looking at the top horizontal part of the sigma. The camera that is located further away is used in our experiments since the planar approximation we make is more valid in that case. The segment of the sigma next to the top horizontal part is used as a probe. This segment is at an angle $33°$ to the horizontal part. Fig. 3.7(b) shows the probe and gallery segments. A few images from the NIST database are shown in Fig. 3.6. The method described before was used to estimate the walking direction and it turned out to be very close to the true value. The synthesis was done as explained above. Some of the results of the synthesis are shown in Fig. 3.6. In order to do gait recognition, we used the fusion

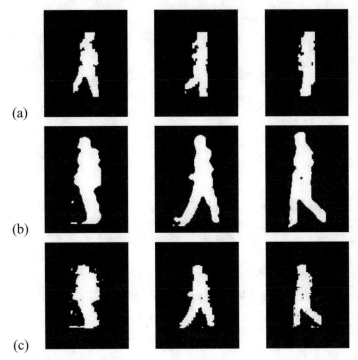

(a)

(b)

(c)

FIGURE 3.6: Examples for the NIST database. (a) Gallery images of person walking parallel to the camera; (b) unnormalized images of person walking at 33° to the camera; (c) synthesized images for (b).

of the leg-dynamics with the height as the cue. This is based on anecdotal evidence that the lower part of the body contains most of the information for gait. The gait recognition result is shown in Fig. 3.7(a). As can be seen the recognition rate is about 60% and that the recognition goes to 100% within six ranks. The performance is significantly better than without the synthesis.

The CMU MoBo database (59) consists of 25 people walking on a treadmill in the CMU 3D room. There are six synchronized cameras evenly distributed around the subject walking on the treadmill. For testing our algorithm we considered views of the slow walk sequences from cameras 3 and 13. Camera 3 captures the exact side view while the camera 13 captures a noncanonical view of the subjects. The sequence from camera 3 was used as the gallery while the sequence from

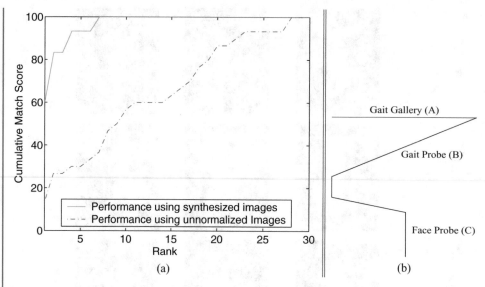

FIGURE 3.7: (a) Gait recognition performance on the NIST database. (b) Σ pattern of path in the NIST database. The face probe sign is for Fig. 5.6.

camera 13 was used as the probe. Since there is no actual translation involved in this case, it is not possible to estimate the image plane angle α or the focal length f. Hence in order to obtain the synthesized images for this database, we use an alternate approach. We considered several points $S = \{(x_1, y_1), \cdots, (x_n, y_n)\}$ on the side of the treadmill, which is a planar rectangular surface. We then constructed the view of this rectangular patch as it would appear in the canonical view. A set of points $S' = \{(x_1', y_1'), \cdots, (x_n', y_n')\}$ is considered on this hypothesized patch. A homography of the nature of (3.23) is estimated using the sets S and S' (65). This homography is then used to obtain the synthesized images for the person. Gait recognition performance using the unsynthesized and synthesized images is shown in Fig. 3.8. Again we found that synthesis results in better gait recognition performance and that using the leg region alone achieves better performance as compared to using the entire body. The true person is identified within the top six matches. For this data set, the assumptions of planarity are not strictly valid. Even then we find significant advantages in the view synthesis approach.

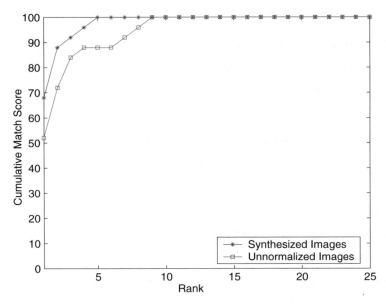

FIGURE 3.8: Cumulative match characteristics for original and synthesized images for the CMU database.

3.3.4 Visual Hull Based View Invariant Approaches

A visual hull (VH) of an object is the intersection of all the extruded cone-like shapes that result from back-projecting the silhouettes in all views (93). Hence, VH can be obtained by volume carving. It is possible to reduce the computation of VH to 2D operations since it contains only points that project onto the silhouettes. Image based visual hull (IBVH) (107) is an efficient geometrically-valid pixel re-projection method to compute the VH. For each pixel in the desired view, the epipolar line in each input view is intersected with the contour approximation, then the intersected 2D line segment is projected back to 3D space to form the VH. The algorithm is able to render a desired view of n^2 pixels in $O(kn^2)$, where k is the number of input views. IBVH is a view-dependent algorithm. It ensures the correctness of the generated image for the desired viewpoint (with the epipolar constraints), with no need to explicitly build the VH in 3D space. After the VH is constructed, its surface is texture mapped using the weighted sum of intensity values in the input images

(106). Considering the visibility during the texture mapping process, an occlusion-compatible warping ordering scheme (110) is used to solve the object occlusion and nonocclusion problem. An advantage of the IBVH technique is its ability to find a balance between accuracy and efficiency. With the widely positioned views as inputs, IBVH allows us to produce the virtual view without finding the wide baseline correspondence. It also provides information about the object's 3D shape and location. Besides, since the VH is formed by volume carving, the noise from input images is greatly reduced in the intersecting process.

Face and gait can be used as multimodal biometric information for human identification. Usually face recognition needs the frontal view of the human face, while gait recognition requires the side view of the human silhouette. With view synthesis technique, both the requirements can be satisfied if we are given multi-view video sequences. Hence face and gait information can be combined to achieve a higher recognition rate. Some promising results have been reported for integrated gait and face recognition using IBVH technique in (61) where a constant velocity Kalman filter is used to estimate the person's motion trajectory and the virtual camera is placed accordingly. This approach will not work if the motion trajectory is hard to estimate, or not available (e.g., turning around). In order to solve the general motion case, we propose to actively move the virtual camera on the view sphere of the object to generate a turntable image collection, and then use a template matching method to select the desired viewpoint for recognition applications. The view sphere of an object is a sphere that is centered at the object and has a fixed radius (7). The turntable image collection is a collection of the object's images that are captured by a camera moving around the object, with the optical axis parallel to the plane that the object stands on. By moving the virtual camera along a properly selected circular trajectory on the view sphere, the turntable image collection can be rendered quickly and efficiently using the IBVH technique. We derive a method by aligning the camera calibration coordinate system and the world coordinate system if they do not coincide each other, the virtual camera's position on the circular trajectory can be decided accurately. Considering the nonrigid free-form human movement, a template matching method and a turning angle comparison method are used to select the desired viewpoint upon the synthesized image collection and generate the pose-invariant views.

All the shape from silhouettes methods are not able to reconstruct concave regions because they are not observable in any silhouette. In (93), it was stated that the VH of an object depends not only on the object itself but also on the region allowed to the viewpoint. The *external visual hull* is related to the convex hull, and the *internal visual hull* can not be observed from any viewpoint outside the convex hull. Observing that in many cases the concave human posture is formed due to the position of arms, we are inspired to explore the possibility of body part based view synthesis with IBVH. Examples of image rendering using IBVH are shown in Figs. 3.9 and 3.10. Several methods have been proposed for human body part segmentation from silhouette (contour). The work in (63) gives a silhouette-based human body labelling template by using topological order-constraints of body parts for different postures. A contour-based body part localization method was presented in (190) with a probabilistic similarity measure which combines the local shape and global relationship constraints to guide body part identification.

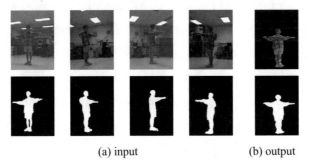

(a) input   (b) output

FIGURE 3.9: (a) An example of IBVH: the images observed from the 4 static cameras (top) and corresponding silhouette images (bottom). (b) The rendered image corresponding to a novel view with texture (top) and without texture (bottom) obtained with IBVH.

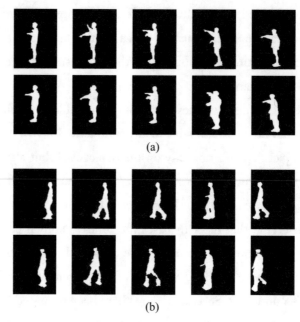

(a)

(b)

FIGURE 3.10: (a) The side view ground truth for the Keck lab sequence (top) and the rendered side view for the Keck lab sequence (bottom). (b) The side view ground truth for the MIT sequence (top) and the rendered side view for the MIT sequence (bottom).

More recently, a hierarchical model fitting method to estimate the 3D shape with density fields was proposed in (18). The body parts of the human can be described accurately with the estimated parameters. We use the work in (190) for body part segmentation because of its simplicity and robustness, where the short-cut rule and the saliency requirement are combined to constrain the other end of a cut, and several computationally efficient strategies are used to reduce the effects of noise. Using this method, the silhouette image in each input view is partitioned into arms and torso (with legs) so that each human part is a convex object. All the parts are separately processed with IBVH, and assembled together to get the final result (see Fig. 3.11). It is possible that the final view has some disconnected or squeezed regions since it is obtained by assembling the separately processed body parts. To prevent this problem, a silhouette image for the desired viewing direction is first generated without segmenting the body parts.

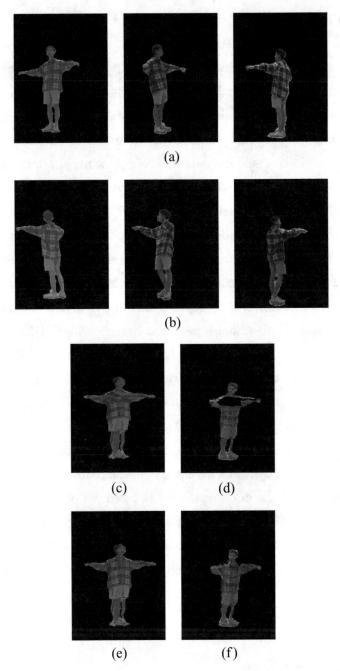

FIGURE 3.11: Two examples of view synthesis of articulating humans with visual hull. (a) and (b) are the input views. (c) and (d) are the texture maping result without using body part segmentation method. (e) and (f) are the texture maping result using body part segmentation segmentation method.

3.4 ROLE OF SHAPE AND DYNAMICS IN GAIT RECOGNITION

In the previous sections, we provided a broad overview of a number of techniques and described a few of them in detail. In spite of the development of methods that give reasonably good results under certain circumstances, there is little understanding of the underlying physical processes that contribute to the gait of a person. It is widely accepted that shape and dynamics contribute to gait. But what is the relative importance of these two cues separately? Physchophysicists have studied this question and have arrived at their own conclusions, some of which we described in Section 3.1. In this section, we describe some of our work in understanding the role of these two cues. Our studies are done from the perspective of a computer scientist and evaluate the role of these cues in computer vision methods for recognition.

In our experiments, shape is represented using Kendall's definition of shape (40). Gait recognition algorithm is described by computing the distance between two sequences of shapes that lie on a spherical manifold. The HMM and DTW methods are used for computing shape similarity, but the distances are computed on the manifold rather than in Euclidean space. Dynamics is represented using an autoregressive moving average (ARMA) model. Our conclusions show that shape plays a role that is more significant than dynamics in human identification using gait. We also note that dynamics has more significance for the problem of activity classification than for the problem of activity based person identification. However, even for the problem of activity classification introducing shape information leads to better classification. These conclusions also allow us to explain the relative performance of various existing methods in computer-based human activity modeling.

3.4.1 Shape Analysis

3.4.1.1 Definition of Shape

"Shape is all the geometric information that remains when location, scale and rotational effects are filtered out from the object."(40). Kendall's representation of shape describes the shape configuration of k landmark points in an m-dimensional space as a $k \times m$ matrix containing the coordinates of the landmarks. In our analysis we have a 2D space and therefore it is convenient to describe the shape vector as a k-dimensional complex vector.

The binarized silhouette of a walking person is obtained. A shape feature is extracted from the binarized silhouette. This feature vector must be invariant to translation because identity should be independent of the location of the individual. It should also be invariant to scaling since identity should not depend on the distance of the subject from the camera. So any feature vector that we obtain must be invariant to translation and scale. This yields the preshape of the walking person in each frame. Preshape is the geometric information that remains when location and scale effects are filtered out. Let the configuration of a set of k landmark points be given by a k-dimensional complex vector containing the position of the landmarks. Let us denote this configuration as X. Centered preshape is obtained by subtracting the mean from the configuration and then scaling to norm one. The centered preshape is given by

$$Z_c = \frac{CX}{\| CX \|}, \qquad C = I_k - \frac{1}{k}1_k 1_k^T, \qquad (3.24)$$

where I_k is a $k \times k$ identity matrix and 1_k is a k-dimensional vector of ones.

3.4.1.2 Distance Between Shapes

The preshape vector that is extracted by the method described above lies on a spherical manifold. Therefore a concept of distance between two shapes must include the non-euclidean nature of the shape space. Several distance metrics have been defined in (40). When the shapes are very close to each other there is hardly any difference between the various shape distances. Since the shapes of a single individual lie close to each other, we will show results using the partial Procrustes distance.

Consider two complex configurations X and Y with corresponding corresponding preshapes α and β. The partial Procrustes distance between configurations

X and Y is obtained by matching their respective preshapes α and β as closely as possible over rotations, but not scale. So,

$$d_P(X, Y) = \inf_{\Gamma \in SO(m)} \| \beta - \alpha\Gamma \| . \quad (3.25)$$

The rotation angle θ that minimizes the partial Procrustes fit is given by $\Gamma = \arg(|\alpha^*\beta|)$.

3.4.1.3 The Tangent Space

The shape tangent space is a linearization of the spherical shape space around a particular pole. Usually the Procrustes mean shape of a set of similar shapes (Y_i) is chosen as the pole for the tangent space coordinates. The Procrustes mean shape (μ) is obtained by minimizing the sum of squares of full Procrustes distances (see (40) for details) from each shape Y_i to the mean shape, i.e.,

$$\mu = \arg \inf_{\mu} \Sigma d_F^2(Y_i, \mu). \quad (3.26)$$

The preshape formed by k points lie on a $k - 1$ dimensional complex hypersphere of unit radius. If the various shapes in the data are close to each other then these points on the hypersphere will also lie close to each other. The Procrustes mean of this data set will also lie close to these points. Therefore the tangent space constructed with the Procrustes mean shape as the pole is an approximate linear space for this data. The Euclidean distance in this tangent space is a good approximation to the various Procrustes distances d_F, d_P, and ρ in shape space in the vicinity of the pole. The advantage of the tangent space is that this space is Euclidean. The Procrustes tangent coordinates of a preshape α is given by

$$v(\alpha, \mu) = \alpha\alpha^*\mu - \mu|\alpha^*\mu|^2, \quad (3.27)$$

where μ is the Procrustes mean shape of the data.

3.4.1.4 Shape-Based Methods for Recognition

In order to study the importance of shape for recognition we use the Kendall's shape vector as the image feature. Recognition performance is obtained on three independent systems: a stance correlation based method, DTW (80), and another method based on the generic HMM framework suggested in (162). The stance correlation method uses only shape cues for recognition. The DTW and HMM attempt to exploit the dynamics in gait, although the dynamics is not explicitly modeled or captured.

1.   Stance correlation: Given a sequence of frames of a person walking, each frame is classified as belonging to one of six stances by first locating the cycle boundaries and then locating the stance boundaries. An exemplar is obtained for each stance as an average of all the frames belonging to that particular stance. The correlation between the corresponding exemplars of two sequences is used as the similarity score between these two sequences. Recognition is performed based on these similarity scores.

2.   Dynamic time warping in shape space: This is a shape-based recognition algorithm using DTW on the spherical shape manifold. Given a sequence of a person walking, the shape of the silhouette is extracted for every frame. We use DTW to obtain the similarity score between two such shape sequences. One major difference from the approach in (80) is that we do the DTW between shapes on a spherical manifold instead of the DTW using Euclidean distances. The Procrustes shape distances are used as the local distance measure.

3.   Hidden Markov model with shape cues: The shape of the silhouette of the walking person is extracted for each frame and is used as the feature vector for the HMM, as described in Section 3.2.

3.4.2   Dynamical Models for Gait

Gait is an activity with structured dynamics associated with it. Therefore, a purely shape based method is not an optimal approach to gait recognition. What role does dynamics play computer recognition algorithms? DTW and HMM attempt to capture the dynamics of gait, though not explicitly. In this section, we consider two dynamical models for gait and provide recognition results using these dynamical models.

3.4.2.1 Stance-Based AR Model

The dynamical model must be insensitive to shape, i.e., if two people with vastly different shapes display the same dynamics, then the model inferred for these two individuals must be the same. This will ensure that we examine the effect of dynamics alone on recognition performance.

The video sequence of a person walking is divided into N distinct parts or stances. Each frame is labeled as belonging to one of N stances. Within each stance, we learn the dynamics of the shape vector as a function of time. The time series of the tangent space projections of the shape vector of each stance is modeled as a Gauss Markov process, i.e.,

$$\underline{\alpha}_j(t) = A_j \underline{\alpha}_j(t-1) + w(t), \qquad (3.28)$$

where w is a zero mean white Gaussian noise process and A_j is the transition matrix corresponding to the jth stance. The parameters $A_j (j = 1, 2, \ldots, N)$ represent the dynamics of that particular sequence. For convenience and simplicity A_j is assumed to be a diagonal matrix. Note that A_j is computed for each stance separately. Since the gait signal is periodic, we would expect at least a second order model. However, for each stance there is no periodicity; hence the first order model is valid. Let us call the average shape of each stance as the exemplar of that stance and represent it as $E_j, j = 1, 2, \ldots, N$. We could, of course, use both the exemplars themselves and the transition matrices for each stance as the model of gait and use them for recognition. But, the exemplars contain shape information. Therefore, we use only the transition matrices for recognition in order to understand the role of dynamics.

For all the sequences in the gallery the transition matrices are obtained and stored. Given a probe sequence, the transition matrix for the probe sequence is computed. The distance between the corresponding transition matrices are added to obtain a measure of the distance between the dynamical models. If A_j and B_j (for $j = 1, 2, \ldots, N$) represent the transition matrices for two sequences, then the distance between models is defined as $D(A, B)$:

$$D(A, B) = \sum_{j=1}^{j=N} ||A_j - B_j||_{\mathrm{F}}, \tag{3.29}$$

where $||_{\mathrm{F}}$ denotes the Frobenius norm. The model in the gallery that is closest to the model of the given probe is chosen as the identity of the person.

3.4.2.2 Linear Dynamical Model

The second approach uses a linear dynamical system to model gait. The problem of gait recognition is transformed to one of learning a dynamical model from the observations and computing the distances between the dynamical models thus learnt. The dynamical model is a continuous state, discrete time model. Since the parameters of the models lie in a non-euclidean space, the distance computations between models is nontrivial. Let us assume that the time-series of shapes is given by $\alpha(t)$, $t = 1, 2, \ldots, \tau$. Then the linear dynamical (or ARMA) model is defined by (12)

$$\alpha(t) = Cx(t) + w(t); w(t) \sim N(0, R) \tag{3.30}$$
$$x(t + 1) = Ax(t) + v(t); v(t) \sim N(0, Q). \tag{3.31}$$

Also, let the cross correlation between w and v be given by S. The parameters of the model are given by the transition matrix A and the state matrix C. We note that the choice of matrices A, C, R, Q, S is not unique. But, we also know (121) that we can transform this model to the "innovation representation." The model parameters in the innovation representation are unique.

We use the tools from the system identification literature to estimate the model parameters. The algorithm is described in (121) and (155). Given observations $\alpha(1), \alpha(2), \ldots, \alpha(\tau)$, we have to learn the parameters of the innovation representation given by $\hat{A}$, $\hat{C}$, and $\hat{K}$. Note that in the innovation representation, the state covariance matrix $\lim_{t\to\infty} E[x(t)x^T(t)]$ is asymptotically diagonal. Let

$[\alpha(1)\alpha(2)\alpha(3)\ldots\alpha(\tau)] = U\Sigma V^T$ be the singular value decomposition of the data. Then

$$\hat{C}(\tau) = U \tag{3.32}$$
$$\hat{A} = \Sigma V^T D_1 V (V^T D_2 V)^{-1} \Sigma^{-1}, \tag{3.33}$$

where $D_1 = [0\ 0; I_{\tau-1}\ 0]$ and $D_2 = [I_{\tau-1}\ 0; 0\ 0]$.

Subspace angles (56) between two ARMA models are defined as the principal angles $(\theta_i, i = 1, 2, \ldots, n)$ between the column spaces generated by the observability spaces of the two models extended with the observability matrices of the inverse models (30). The subspace angles between two ARMA models ($[A_1, C_1, K_1]$ and $[A_2, C_2, K_2]$ can be computed by the method described in (30). Using these subspace angles $\theta_i, i = 1, 2, \ldots n$, Martin distance($d_M$), gap distance($d_g$) and Frobenius distance(d_F) between the ARMA models are defined as follows:

$$d_M^2 = \ln \prod_{i=1}^{n} \frac{1}{\cos^2(\theta_i)} \tag{3.34}$$
$$d_g = \sin \theta_{\max} \tag{3.35}$$
$$d_F^2 = 2 \sum_{i=1}^{n} \sin^2 \theta_i. \tag{3.36}$$

The various distance measures do not alter the results significantly. We show the results using the Frobenius distance.

3.4.3　Experimental Results

We conducted recognition experiments using all the five methods: stance correlation, DTW on shape space, HMM using shape, stance based AR, and linear dynamical system on the USF data set. The CMS curves for challenge experiments A–G are shown. An analysis of the various CMS curves is also provided.

The following conclusions can be drawn from Fig. 3.12:

- The CMS curves of the stance correlation method show that shape without any dynamic cues provides recognition performance below baseline.

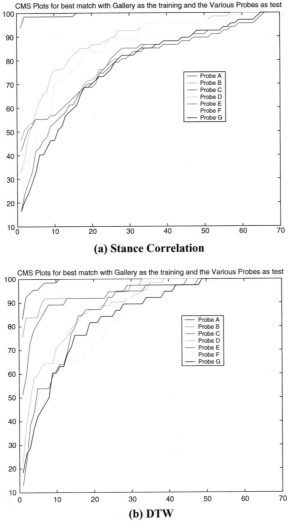

(a) Stance Correlation

(b) DTW

FIGURE 3.12: CMS curves of experiments A–G using (a) stance correlation on exemplars and (b) DTW on image shape.

- The CMS curves of the DTW method are better than those of stance correlation and closer to baseline.

- The improvement in the CMS curves in the DTW over that of the stance correlation method can be attributed to the presence of this implicit dynamics, because the algorithm tries to synchronize two warping paths.

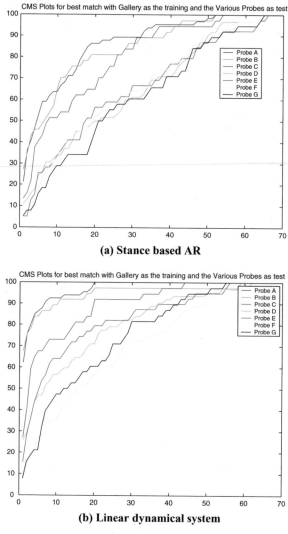

FIGURE 3.13: CMS curves of experiments A–G using (a) stance-based AR and (b) linear dynamical system models.

- The performance of the HMM with shape as the feature vector is very similar to that of the DTW.

The conclusions that can be drawn from Figure 3.13 are as follows:

- Both methods based on dynamics do not perform as well as the methods based on shape.

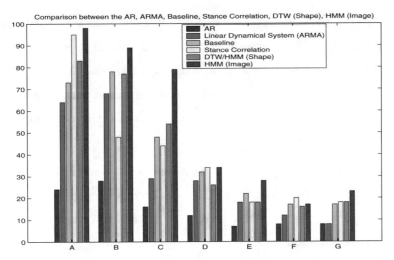

FIGURE 3.14: Bar diagram comparing the identification rate at rank 1 of various algorithms.

- The results support our belief that dynamics helps to boost recognition performance but is not sufficient as a stand-alone feature for recognition.

- The recognition performance of the linear dynamical system is better than that of the stance-based AR model. Experiments also suggest that the observation matrix (matrix containing the eigen vectors: C) contributes significantly in this recognition performance, while the role the transition matrix (A) plays is minimal.

Fig. 3.14 shows a comparison of the performance of the various shape and kinematics-based algorithms. It is clearly seen that shape based algorithms perform better than purely dynamics-based algorithms. HMM-based algorithms, which try to combine both shape and dynamics, perform best. Moreover, the above experiments show that shape plays a more important role in existing computer based recognition algorithms.

3.5   CONCLUSIONS

We briefly summarize the main developments over the last few years in the area of human identification using gait:

- Most of the successful methods in gait recognition are appearance based (162; 169). There have been a few model-based approaches, but their recognition performance on large data sets has been lower (96; 167). This may be because of errors in estimation of model parameters. However, such methods can be useful for other kinds of applications, like detecting an abnormal gait.

- Most methods on gait recognition assume a side view of the person is available. Some of the techniques that have been developed for view-invariant recognition are the VH approach in (61) and the planar SfM approach described in Section 3.3 (79).

- The effect of different parameters in gait recognition from video is not yet very well understood. The effect of cadence and stride length have been investigated in (17). We have tried to analyze the effect of shape and dynamics in the process and the results of this were presented in Section 3.4.

- The GaitChallenge Data set (128; 129) has emerged as a standard testbed for new recognition algorithms, since it captures the variations of a large number of covariates for a large group of people.

CHAPTER 4

Human Activity Recognition

Activity modeling, recognition, and anomaly detection from video sequences have applications in video surveillance and monitoring, human computer interaction, video transmission and analysis, medicine, computer graphics, and virtual reality. Various techniques have been used for the study of actions from sequences of images. A review of some of the literature in activity modeling and recognition is provided in a subsequent section.

In order to recognize different activities, it is necessary to construct an ontology of various normal (both frequent and rare) events. Deviations from a preconstructed dictionary can then be classified as abnormal events. It is also necessary that the representation be invariant to the viewing direction of the camera and independent of the number of cameras (i.e. should be scalable to a video sensor network). Trajectories, usually computed from 2D video data, are a natural starting point for activity recognition systems. Trajectories contain a lot of information about the underlying event that they represent. However, one must do more than track a set of points over a sequence of images and infer about the event from the set of tracks. Trajectories are ambiguous (different events can have the same trajectory) and depend on the viewing direction. Also, identifying events from trajectories requires the enunciation of a set of heuristics, which can vary from one instance to another of the same event. Hence, it is important to have a *proper* intermediate step in the leap from trajectories to event models (see Fig. 4.1). In a recent paper, Rao *et al.*

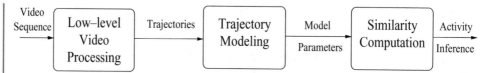

FIGURE 4.1: The framework for activity inference. We start by computing trajectories from a video sequence, then fit models to these trajectories (e.g. dynamic instants, Kendall's shape, or 3D models as proposed here), and finally compute the similarity between model parameters for inferring about the activity.

(135) proposed a method for representing a trajectory in terms of dramatic changes in its speed and direction. They represented a human activity in terms of action units called dynamic instants and intervals, and their method was motivated by studies on human perception. In (176), the authors proposed a shape model (along the lines of Kendall's shape theory) on the set of points in each image frame and described an activity by the dynamics of the shape. In this chapter, we will present two methods that use 2D and 3D shape models for each activity (142; 176; 177) and present recognition results.

4.1 LITERATURE REVIEW

Event analysis from video sequences has a long history in the computer vision literature. We provide a brief review of past work dealing with the general problem of event recognition as well as the special situation of human activity analysis and inference.

Most of the early work on activity representation comes from the field of artificial intelligence (AI). One of the earliest attempts at developing a general scheme for representing activities and building a system based on it was reported by Tsuji *et al.* (172). They applied their principles to understanding activities taking place in simple cartoon films. Neumann and Novak (119) proposed a hierarchical representation of event models, with each model being a template that can be matched with scene data. Natural language descriptions of activities can be mapped on this hierarchical model. More recent work comes from the fields of image understanding and visual surveillance. The formalisms that have been employed include HMMs, logic programming, and stochastic grammars. Nagel proposed (117) an early approach for obtaining conceptual descriptions from image sequences, which could then be used for representing and recognizing activities. Dousson *et al.* (39), Kuniyoshi and Inoue (90), and Buxton and Gong (25) presented models and algorithms for situation analysis from video data. Davis and Bobick (37) developed a scheme for characterizing human actions based on the concept of "temporal templates." Bremond and Thonnat (23) investigated the use of contextual information in activity recognition. The use of declarative models for activity recognition from video sequences was described in (139). Each activity was represented by a set of conditions between different objects in the scene. This translated into a constraint satisfaction problem for recognizing activity. A method for representing a scenario by a set of subscenarios and constraints combining these subscenarios was proposed in (179). Castel *et al.* (27) have developed a system for high-level interpretation of image sequences, in which they clearly separate the numerical and symbolic levels of representation and reasoning. Over the last 5 years, HMMs (158; 183) have been used for recognizing American Sign Language and parametric gestures respectively.

One of the requirements of any reliable recognition scheme is the ability to handle uncertainty. Many uncertainty-reasoning models have been actively pursued in the AI and image understanding literature, including belief networks (125) and Dempster–Shafer theory (148). A method for inferring activities of humans and vehicles in airborne video using dynamic Bayesian networks was proposed in (4). Large belief networks (BNs) have been used in several video interpretation applications. For example, Intille and Bobick (72) have used large BNs to classify football plays. A system for classifying human motion and simple human interactions using small BNs was developed by Remagnino *et al.* (136). A method of generating high-level descriptions of traffic scenes was implemented by Huang *et al.* (71) using a dynamic Bayes' network (DBN). In (58), the authors proposed building a tracking and monitoring system using a "forest of sensors" distributed around the site of interest. Their approach involved tracking objects in the site, learning typical motion and object representation parameters (e.g. size and shape) from extended observation periods and detecting unusual events in the site. In (68), the authors proposed a method for recognizing events involving multiple objects using Bayesian inference. Kendall's shape theory was used to model the interactions of a group of people and objects in (176).

A specific area of research within the broad domain of event recognition is human motion modeling and analysis. The ability to recognize and track human activity using vision is one of the key challenges that must be overcome before a machine is able to interact meaningfully with a human-inhabited environment. Traditionally, there has been a keen interest in studying human motion in various disciplines. In psychology, Johansson conducted classic experiments by attaching light displays to various body parts and showed that humans can identify motion when presented with only a small set of these moving dots (76). Muybridge captured the first photographic recordings of humans and animals in motion in his famous publication on animal locomotion toward the end of the nineteenth century (116). In kinesiology the goal has been to develop models of the human body that explain how it functions mechanically (64). The challenge to the computer vision community is to devise efficient methods to automatically track moving humans in a video sequence, reconstruct nonrigid 3D models, and infer about the various activities being performed by the subjects. A survey of some of the earlier methods used in vision for tracking human movement can be found in (53). In a more recent work, an activity recognition algorithm using dynamic instants was proposed in (135). In (123; 124), each human action was represented by a set of 3D curves, which are quasi-invariant to the viewing direction.

The various methods can be classified as either 2D or 3D approaches. Two-dimensional approaches are effective for applications where precise pose recovery is not needed or possible due to low image resolution (e.g. tracking pedestrians in a surveillance setting). However, it is unlikely that they will perform well in applications that require a high level of discrimination between various unconstrained and complex human movements (e.g. humans making gestures while walking, social interactions, dancing etc.). In such applications, 3D approaches are preferred because they can recover body pose, which allows better prediction and handling of occlusion and collision.

4.2 ACTIVITY MODELING USING KENDALL'S SHAPE THEORY

Kendall's statistical shape theory (40; 84) (described in Section 3.4) has been used to model the shape of the configuration of a group of moving objects and its deformations over time. The notion of separating the motion of a deforming shape into motion of an average shape and its deformations described by Soatto and Yezzi in (156) can be extended to "shape activities." A "static shape activity" as one in which the average shape formed by the moving points remains constant with time and the deformation process is stationary. A "dynamic shape activity" on the other hand has a time varying average shape and/or a nonstationary shape deformation process.

Kendall's shape analysis methods describe the shape of a fixed number of landmarks and so when the number of point objects is not fixed with time, we resample the curve obtained by connecting the object locations at time t to represent it by a fixed number of points, k. The order in which the object locations are joined is kept the same (shape is not invariant to change in ordering of the points). The complex vector formed by these k points (x and y coordinate forming the real and imaginary parts) is then centered using Eq. (3.24) to give the observation vector sequence, $\{Y_t\}$. We first assume that handpicked or accurately measured object location data is available (negligible observation noise). The observation vector is normalized for scale (to obtain the preshape) and generalized Procrustes analysis is performed on this sequence of preshapes to obtain the Procrustes mean shape, μ. The preshapes are aligned to μ and tangent coordinates at μ evaluated using Eq. (3.27). The complex tangent coordinate vector is rewritten as a real vector of twice the complex dimension.

4.2.1   Shape Dynamics in Tangent Space

Let the vector of tangent coordinates be represented by $v_t \in \mathcal{R}^{2k-4}$. The origin of the tangent hyperplane is chosen to be the tangent coordinate of μ and hence the data projected in tangent space has zero mean by construction. The time correlation between the tangent coordinates is learnt by fitting a stationary *Gauss–Markov model* as described in our earlier work (175; 176), i.e.

$$E[v_t] = 0$$
$$v_t = Av_{t-1} + n_t, \tag{4.1}$$

where $n_t{}^1$ is a zero mean i.i.d. Gaussian process and is independent of v_{t-1}. The details of evaluating the covariance matrix of v_t, Σ_v, the autoregression matrix A and covariance of noise Σ_n (assuming stationarity and ergodicity) are discussed in (175)

Based on the stationary Gauss–Markov model described above we have,

$$f^0(v_t) \sim \mathcal{N}(0, \Sigma_v), \quad \forall t$$
$$f^0(v_{t+1}|v_t) \sim \mathcal{N}(Av_t, \Sigma_n). \tag{4.2}$$

Thus any $L+1$ length sequence, $\{v_{t-L}, \ldots, v_{t-1}, v_t\}$, will have a joint Gaussian distribution.

4.2.2   Abnormality Detection: Fully Observed Case

We have assumed till now that the noise in the shape of the observations is negligible compared to the system noise, n_t, and hence we have a fully observed dynamical model. For such a test observation sequence, we can evaluate the tangent coordinates (v_t) directly from the observations (Y_t), as described above.

The following hypothesis is used to test for abnormality. A given test sequence is said to be generated by a *normal activity iff* the probability of occurrence of its

---

[1] Note that to simplify notation, we do not distinguish between a random process and its realization in the rest of the chapter.

tangent coordinates using the pdf defined by (4.2) is large (greater than a certain threshold). Thus the distance to activity statistic for an $L + 1$ length observation sequence ending at time t, $d_{L+1}(t)$, is the negative log likelihood (without the constant terms) of the tangent coordinates of the observation, i.e.

$$d_{L+1}(t) = v_{t-L}^T \Sigma_v^{-1} v_{t-L} + \sum_{\tau = t-L+1}^{t} (v_\tau - Av_{\tau-1})^T \Sigma_n^{-1} (v_\tau - Av_{\tau-1}). \qquad (4.3)$$

We test for abnormality at any time t by evaluating $d_{L+1}(t)$ for the past $L + 1$ frames. In the results section, we refer to this as the "log likelihood metric" (even though it is not actually a "metric").

4.2.3 Partially Observed "Shape" Activity Model

When noise in the observations (projected in shape space) is comparable to the system noise, the above model will fail. This is because tangent coordinates estimated directly from this very noisy observation data would be highly erroneous. Observation noise in the point locations will be large in most practical applications especially with low resolution video. In this case, we have to solve the joint problem of *filtering* out the actual configuration (Z_t) and the corresponding shape from the noisy observations ($Y_t = Z_t + w_t$) and also *detecting* abnormality (as a change in shape). Since Z_t is now unknown, so is the corresponding v_t and we thus have a partially observed nonlinear dynamical system (38) with the following system (state transition) and observation model.

The system model includes the shape space dynamics (the Gauss–Markov model on tangent coordinates) and also the dynamics of the scale and rotation.[2] Even though we are interested only in shape dynamics, modeling the rotation and scale dynamics as a first-order stationary process helps to filter out sudden changes in scale/rotation caused by observation noise, which would otherwise get confused

---

[2] In our current implementation, we have not modeled translation dynamics (we use a translation normalized observation vector) assuming that observation noise does not change the centroid location too much.

as sudden changes in shape space. Also, this dynamical model on scale, rotation (or translation) could model random motion of a camera due to its being inside a UAV (unmanned air vehicle) or any other unstable platform. The observation model is a mapping from state space (tangent coordinates for shape, scale, and rotation) back to configuration space, with noise added in configuration space.

4.2.3.1 System Model Under Noisy Observations

The state vector X_t is composed of $X_t = [v_t^T, \theta_t, \mathbf{s}_t]^T$, where v_t are the tangent coordinates of the unknown configuration Z_t, $\theta_t = \arg(Z_t^* \mu)$ is the rotation normalization angle, and $\mathbf{s}_t = ||Z_t||$ is the scale. The transition model for shape (v_t) is discussed in Section 4.2.1. The scale parameter at time t is assumed to follow a Rayleigh[3] distribution about its past value. The rotation angle is modeled by a uniform distribution with the previous angle as the mean. We have

$$v_t = A v_{t-1} + n_t, \quad n_t \sim \mathcal{N}(0, \Sigma_n)$$
$$s_t = r_t s_{t-1}, \quad r_t \sim \text{Rayleigh}(\sqrt{2/\pi})$$
$$\theta_t = \theta_{t-1} + u_t, \quad u_t \sim \text{Unif}(-a, a) \tag{4.4}$$

with initial state distribution

$$v_0 \sim \mathcal{N}(0, \Sigma_v)$$
$$s_0 \sim \text{Rayleigh}(\bar{s}_0)$$
$$\theta_0 \sim \text{Unif}(\bar{\theta}_0 - a_0, \bar{\theta}_0 + a_0). \tag{4.5}$$

The model parameters A, Σ_n, Σ_v are learnt using a single training sequence of a normal activity and assuming stationarity for v_t as described in 4.2.1. The parameter a is learnt as $a = \max_t |\theta_t - \theta_{t-1}|$. Note that in this chapter we have assumed a stationary system model for v_t. But in general, the framework described here is applicable even if Σ_v, A, Σ_n are time varying (nonstationary process).

---

[3] Rayleigh distribution chosen to maintain nonnegativity of the scale parameter.

4.2.3.2 Observation Model

In our current implementation, we assume that independent Gaussian noise with variance σ_{obs}^2 is added to the actual location of the points, i.e.[4]

$$Y_t \sim \mathcal{N}(Z_t, \sigma_{\text{obs}}^2 I_{2k})$$
$$Z_t = h(X_t) = s_t[(1 - v_{t_c}^* v_{t_c})^{1/2}\mu + v_{t_c}], e^{-j\theta_t}, \qquad (4.6)$$

where $h(X_t)$ is the inverse mapping from tangent space to preshape space followed by scaling by s_t (40; 176).

In general, both σ_{obs}^2 and μ can be time varying and the observation noise need not be i.i.d. in all the point object locations. Also, to take care of outliers, one could allow a small probability (p_{out}) of any point j occurring anywhere in the image with equal probability (uniform distribution).

4.2.4 Particle Filter for Noisy Observations

We use the state transition kernel given in (4.4) and the observation likelihood given by (4.6) in the particle filtering framework (38). The PF provides at each time t, an n sample empirical estimate of the distribution of the state at time t given observations upto time $t - 1$ (prediction) and the distribution of the state given observations upto time t, $\pi_{t|t}^n(v_t, s_t, \theta_t)$ (update). For abnormality detection, only the marginal of shape, $\pi_{t|t}^n(v_t)$ is used. The most basic form of particle filtering suffices for our current problem.

4.2.4.1 Abnormality Detection

We test for abnormality based on the following hypothesis. A test sequence of observations, $\{Y_t\}$ is said to be generated by a *normal activity iff*

(a) it is "correctly tracked" by the particle filter trained on the dynamical model learnt for a normal activity. We test this by thresholding the distance between the

---

[4] $v_{t_c} \in \mathcal{C}^{k-2}$ is the complex version of $v_t \in \mathcal{R}^{2k-4}$.

observation and its prediction based on past observations, i.e. for normalcy,

$$(Y_t - E[Y_t|Y_{0:t-1}])^2 = (Y_t - E_{\pi_{t|t}}[h(X_t)])^2 < \gamma. \qquad (4.7)$$

and

(b) the expectation under $\pi_{t|t}(v_t)$ of the negative log-likelihood of normalcy of the tangent coordinates [expectation under $\pi_{t|t}$ of $d_1(t)$ from Equation (4.3)] is below a certain normalcy threshold, η, i.e.

$$E \triangleq E_{\pi_{t|t}}[-\log f^0(v_t)] < \eta. \qquad (4.8)$$

The PF estimate $\pi_{t|t}^n$ will approximate $\pi_{t|t}$ correctly only if the observations are "correctly tracked" by the PF and hence only in the "correctly tracked" case, E can be estimated using the PF distribution. Also, note that E is actually the Kullback Leibler distance between the pdf corresponding to $\pi_{t|t}$ and the normal activity pdf of tangent coordinates, f^0, plus the entropy of $v_t|Y_{0:t}$. Hence this statistic is referred to as the K-L distance in the results section.

The expression for E is approximated by E_n as follows:

$$E_n \triangleq E_{\pi_{t|t}^n}[-\log f^0(v_t)] = \frac{1}{n}\sum_{i=1}^{n} v_t^{(i)^T}\Sigma_v^{-1}v_t^{(i)} + C, \qquad (4.9)$$

where $C \triangleq \log\sqrt{(2\pi)^{2k-4}|\Sigma_v|}$.

Now, a "drastic" abnormality will cause the PF to lose track and hence will get detected using (a). If the abnormality is a 'slow' one (say a person slowly moving away in a wrong direction), the PF will not lose track. But a systematically increasing bias is introduced in the tangent coordinates (they no longer remain zero mean) and hence the expected negative log likelihood of normalcy will be large in this case causing (b) to be violated. Since the PF does not lose track in this case, the PF distribution estimates $\pi_{t|t}$ correctly and hence E can be estimated using a PF in this case.

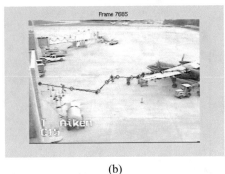

(a)   (b)

FIGURE 4.2: (a) A "normal activity" frame with shape contour superimposed. (b) Contour distorted by spatial abnormality. Note that the normal shape here appears to be almost a straight line, but that is just coincidence; our framework can deal with any kind of polygon formed by the point objects (landmarks). Also, the shape of landmarks does not distinguish open and closed polygons.

4.2.5   Experimental Results

A video sequence of passengers getting out of a plane and walking toward the terminal (Fig. 4.2) was used as an example of "static shape activity" to test our algorithm. Since the number of passengers varies over time, the polygon formed by joining their locations (in the same order always) is resampled to obtain a fixed number of landmarks. We have tested the performance of the algorithm on simulated "spatial" and "temporal" abnormalities (58), since we do not have real sequences with abnormal behavior. Spatial abnormality [shown in Fig. 4.2(b)] is simulated by making one person deviate from his original path. This simulates the case of a person deciding to not walk toward the terminal. Temporal abnormality is simulated by fixing the location of one person thus simulating a stopped person (which can be a suspicious activity too). When the person behind the stopped person goes ahead of him, the loop formed causes the shape to change. Note that since we are using the shape of discrete points, ordering matters and it is for this reason that a stopped person gets detected as an abnormal shape.

We first show results for the case of low (negligible) observation noise, using the log likelihood metric defined in Section 4.2.2. Given a test sequence, at every

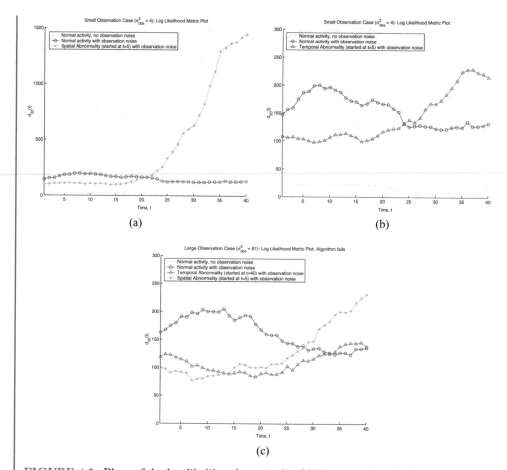

FIGURE 4.3: Plots of the log likelihood metric ($d_{20}(t)$) for normal and abnormal activities. (a) and (b) Comparison of normal activity with spatial and temporal abnormality, respectively, for the case of small observation noise ($\sigma^2_{obs} = 4$). (c) The failure of the algorithm for large observation noise ($\sigma^2_{obs} = 81$). Note that the abnormality was introduced at $t = 5$.

time instant t we apply the log likelihood metric to the past L frames with $L = 20$, i.e. $d_{20}(t) = -\log f^0(v_{t-19}, v_{t-18}, \ldots, v_t)$. Reducing L will detect abnormality faster but will reduce reliability. In Fig. 4.3(a), the cyan dashed line plot is for the case of zero observation noise (handpicked points). The blue circles (o) plot shows the metric for a normal activity with $\sigma^2_{obs} = 4$ ($\sigma = 2$ pixel) Gaussian noise added to the handpicked points, while the green stars (*) plot is for a spatial abnormality (also with the same amount of observation noise) introduced at $t = 5$, for 40 frames. Fig. 4.3(b) shows the same plots for a temporal abnormality (plotted with red triangles). The spatial abnormality gets detected (visually) around $t = 20$, while the temporal one takes a little longer. Some of the lag in both cases is because of $L = 20$. Fig. 4.3(c) shows the same plots but with $\sigma_{obs}^2 = 81$. The metric now confuses normal and abnormal behavior, as discussed in Section 4.2.3.

In Fig. 4.4, we show results for 9 pixel observation noise ($\sigma_{obs}^2 = 81$) but with the observation noise now incorporated into the dynamic model (partially observed dynamic model as discussed in Section 4.2.3). We show plots for the more difficult case of "slow abnormality" where the tracking errors are small even for the abnormal activity. Hence the K-L metric (expected log likelihood) is needed to distinguish between normal and abnormal behavior. Fig. 4.4(a) shows the plot for a spatial abnormality (green stars, *) introduced at $t = 5$, which gets detected around $t = 7$, while as shown in Fig. 4.4(b), the temporal abnormality (red triangles) takes a little longer to get detected. The K-L metric plots for two instances of normal activity with the same amount of noise added are shown in both (a) and (b) with blue circles (∘) and magenta crosses (×).

4.3   ACTIVITY MODELING USING 3D SHAPE

We now discuss a different approach for activity modeling and recognition. The intermediate processing step of Fig. 4.1 is a 3D nonrigid representation of the activity. The 3D representation captures the 3D configuration and dynamics of the set of points taking part in the activity and is independent of the viewing direction of the camera. Also, the method works whether we have a single camera or a network of cameras looking at the scene. The 3D shape estimation is done using the factorization theorem (170), modified for nonrigid shapes (171). A similarity measure between different 3D models is used to classify the various activities. Most existing activity recognition algorithms are appearance based; the factorization method is an attempt at using 3D models for activity classification. New activities that have not been modeled *a priori* can also be identified as such, which in turn leads to the

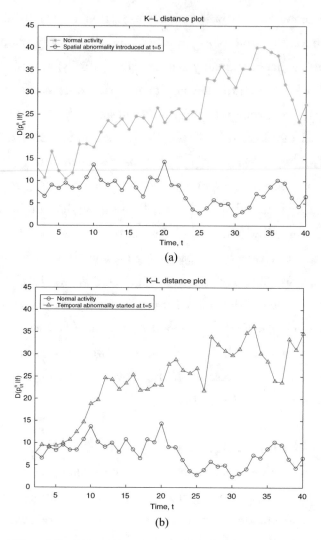

FIGURE 4.4: Plots of the K-L metric which works in large observation noise ($\sigma^2_{\text{obs}} = 81$). (a) and (b) comparison of normal activity with spatial and temporal abnormality, respectively. Note that the abnormality was introduced at $t = 5$.

detection of abnormal activities. Two different kinds of experimental results are presented in order to test the efficacy of our approach. The first set of experiments is done for various activities being carried out by a single individual. We show that we are able to recognize each of those activities. We also demonstrate the view-invariant and multicamera features of this method. In the second experiment, a group of people get off an airplane and are walking toward the terminal (same as in previous section). We model this event and detect any abnormalities that may occur. Finally, we show the application of this approach to the problem of video summarization. We would like to clarify that we do not address the issue of obtaining reliable trajectories in this description, since we consider this to be a separate research problem. A set of heuristics, along with low-level image processing tools, was used to generate reliable tracks in our test video data.

4.3.1   Computing the 3D Models

We hypothesize that each activity can be represented by a linear combination of 3D basis shapes. Difference between the basis shapes can be used to compute the similarity between two activities. Mathematically, if we consider the trajectories of P points taking part in the activity, then the overall configuration of the P points is represented as a linear combination of the basis shapes as

$$S = \sum_{i=1}^{K} l_i S_i, \qquad S, S_i \in \Re^{3 \times P} \quad l \in \Re. \tag{4.10}$$

The choice of K will depend on the particular application and we will explain the details of it when we describe the experiments. A general method for estimating K was proposed in [*A Measure of Deformability of Shapes, With Applications to Human Motion Analysis*, A. Roy-Chowdhury. IEEE Computer Vision and Pattern Recognition, 2005]. We will assume that we have methods to obtain the trajectories accurately. In practice, this was done using a set of heuristics, which we will not describe in detail. Also we will assume a weak perspective projection model for the camera.

A number of methods exist in the computer vision literature for estimating the basis shapes. In (170), the authors considered P points tracked across F frames in order to obtain two $F \times P$ matrices $\mathbf{U}$ and $\mathbf{V}$. Each row of $\mathbf{U}$ contains the x-displacements of all the P points for a specific time frame, and each row of $\mathbf{V}$ contains the corresponding y-displacements. It was shown in (170), that for 3D rigid motion under orthographic camera model, the rank, r, of $\left[\frac{\mathbf{U}}{\mathbf{V}}\right]$ has an upper bound of 3. The rank constraint is derived from the fact that $\left[\frac{\mathbf{U}}{\mathbf{V}}\right]$ can be factored into two matrices

$\mathbf{M}_{2F \times r}$ and $\mathbf{S}_{r \times P}$, corresponding to the pose and 3D structure of the scene, respectively. In (171), it was shown that for nonrigid motion, the above method could be extended to obtain a similar rank constraint, but one that is higher than the bound for the rigid case. We will use the latter method for computing the basis shapes. We will outline the basic steps of their approach in order to clarify the notation.

Given F frames of a video sequence with P moving points, we obtain the trajectories of all these points over the entire video sequence. These P points are represented in a measurement matrix as

$$\mathbf{W}_{2F \times P} = \begin{bmatrix} u_{1,1} & \cdots & u_{1,P} \\ v_{1,1} & \cdots & v_{1,P} \\ \vdots & \vdots & \vdots \\ u_{F,1} & \cdots & u_{F,P} \\ v_{F,1} & \cdots & v_{F,P} \end{bmatrix}, \tag{4.11}$$

where $u_{f,p}$ represents the x-position of the pth point in the fth frame and $v_{m,p}$ represents the y-position of the same point.

Under weak perspective projection, the P points of a configuration in a frame f are projected onto 2D image points $(u_{f,i}, v_{f,i})$ as

$$\begin{bmatrix} u_{f,1} & \cdots & u_{f,P} \\ v_{f,1} & \cdots & v_{f,P} \end{bmatrix} = \mathsf{R}_f \left(\sum_{i=1}^{K} l_{f,i} S_i \right) + \mathbf{T}_f, \tag{4.12}$$

where

$$\mathsf{R}_f = \begin{bmatrix} r_1 & r_2 & r_3 \\ r_4 & r_5 & r_6 \end{bmatrix} \triangleq \begin{bmatrix} \mathsf{R}_f^{(1)} \\ \mathsf{R}_f^{(2)} \end{bmatrix}. \tag{4.13}$$

R_f represents the first two rows of the full 3D camera rotation matrix and $\mathbf{T}_f$ is the camera translation. The translation component can be eliminated by subtracting out the mean of all the 2D points, as in (170). We now form the measurement matrix $\mathbf{W}$, which was represented in (4.11), with the means of each of the rows subtracted. The weak perspective scaling factor is implicitly coded in the configuration weights, $\{l_{f,i}\}$.

Using (4.11) and (4.12), it is now easy to show that

$$\mathbf{W} = \begin{bmatrix} l_{1,1}\mathbf{R}_1 & \cdots & l_{1,K}\mathbf{R}_1 \\ l_{2,1}\mathbf{R}_2 & \cdots & l_{2,K}\mathbf{R}_2 \\ \vdots & \vdots & \vdots \\ l_{F,1}\mathbf{R}_F & \cdots & l_{F,K}\mathbf{R}_F \end{bmatrix} \begin{bmatrix} S_1 \\ S_2 \\ \vdots \\ S_K \end{bmatrix} \tag{4.14}$$

$$= \mathbf{Q}_{2F \times 3K} \cdot \mathbf{B}_{3K \times P}. \tag{4.15}$$

The matrix $\mathbf{Q}$ contains the pose for each frame of the video sequence and the weights $l_1, \ldots, l_K$. The matrix $\mathbf{B}$ contains the basis shapes corresponding to each of the activities. In (171), it was shown that $\mathbf{Q}$ and $\mathbf{B}$ can be obtained using singular value decomposition (SVD) as $\mathbf{W}_{2M \times P} = \mathbf{U}\mathbf{D}\mathbf{V}^T$ and $\mathbf{Q} = \mathbf{U}\mathbf{D}^{\frac{1}{2}}$ and $\mathbf{B} = \mathbf{D}^{\frac{1}{2}}\mathbf{V}^T$.

4.3.2 Activity Inference

Having obtained the 3D models, the next step is to classify the various activities. Our approach for activity inference consists of a learning/training phase and a testing one. During the training phase, the 3D models for various activities are computed. Given a test sequence, the 3D model estimated from this sequence is compared with that learned before and a similarity score is computed based on a measure of the difference of the two 3D models. The exact method for computing this difference is based on the particular application. Our experiments are concerned with two different kinds of activities. In the first, we will consider different activities performed by a single individual. In this case the 3D model of the person conducting the activity is computed at each instant of time. The difference between two 3D models is computed using joint angles. In the second case, we consider the activity of a group of objects, each object being represented as a point. Specifically, we consider an airport surveillance scenario where there are two classes of objects: passengers and vehicles. The paths followed by the passengers and the vehicles are very different, and the 3D model of these paths are estimated and compared. The values of the weighting coefficients, l_i, are used to compute the 3D models. Details of the processes are available in the following section.

4.3.3   Experimental Results

4.3.3.1 Human Action Recognition

We used our method to classify the various activities performed by an individual. We used the motion-capture (MOCAP) data available from Credo Interactive, Inc. and Carnegie Mellon University in the BioVision Hierarchy and Acclaim formats. The combined data set included a number of subjects performing various activities, like walking, jogging, sitting, crawling, brooming, etc. For each of these activities, we had multiple video sequences. Also, many of the activities contained video from different viewpoints.

Using the video sequences and the theory outlined above, we compute the basis shapes and their combination coefficients [see Eq. (4.10)]. We found that the first basis shape, S_1, contained most of the information. The first basis shapes and the combination coefficients, l, plotted across time are shown in Fig. 4.5 for

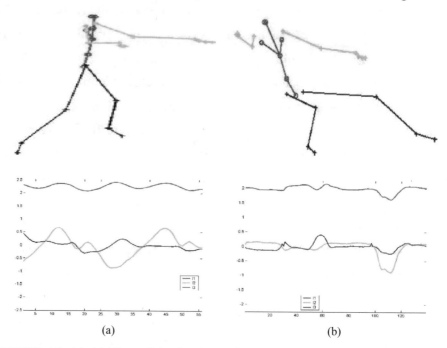

(a)   (b)

FIGURE 4.5: (a)–(c) Plots of the first basis shape, S_1, and combination coefficients l_i (against times) for walk, sit, and broom sequences, respectively. (d)–(f) Plots of the first basis shape, S_1, and combination coefficients l_i (against times) for jog, blind walk, and crawl sequences, respectively.

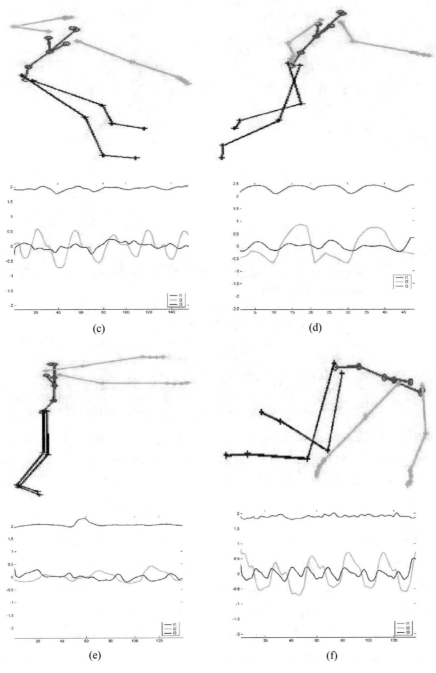

(c)

(d)

(e)

(f)

FIGURE 4.5: (*Cont.*)

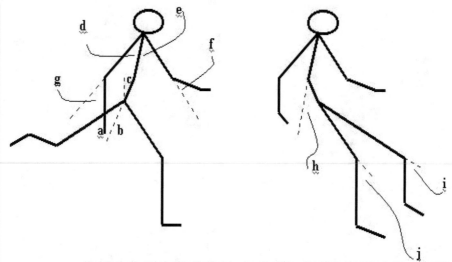

Angles we are using in our correlation criteria (ordered from highest weighted to lowest):

1. c -> angle btw. Hip-abdomen and vertical y axis
2. h -> angle btw. Hip-abdomen-chest
3. (a+b)/2 -> average angle btw. two legs and abdomen-hip axis
4. (b-a)/2 -> the angle difference between two upper legs
5. (i+j)/2 -> average angle btw. upper legs and lower legs
6. (d+e)/2 -> average angle btw. upper arms and abdomen-chest axis
7. (f+g)/2 -> average angle btw. upper arms and lower arms

FIGURE 4.6: The various angles used for computing the similarity of two models are shown. The text below describes the seven dimensional vector computed from each model and whose correlation determines the similarity scores.

six different activities. Since the values of l_i are small for $i > 1$, we used only the first basis shape to compute the similarity between the various activities. In order to compute the similarity, we considered the various joint angles between the different parts of the estimated 3D models. The angles considered are shown in Fig. 4.6. The idea of considering joint angles for activity modeling has been used before, e.g. in gait recognition (167). We considered the seven dimensional vector obtained from the angles as shown in Fig. 4.6. The correlation between two such angle vectors was used as the measure of similarity. Any other combination, or even a

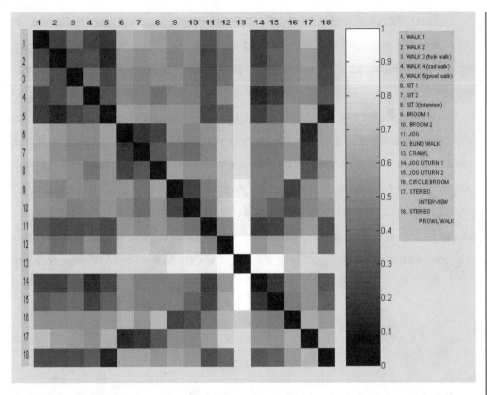

FIGURE 4.7: The similarity matrix for the various activities, including ones with different viewing directions and multiple cameras.

different representation altogether, could have been used for this purpose. However, we found this measure had an intuitive appeal based on the physics of human motion.

The similarity matrix is shown in Fig. 4.7. For the moment, consider the upper 13 × 13 block of this matrix. We find that the different walk sequences are close to each other. Similar observations can be made for the sitting and brooming sequences. The jog sequence, besides being closest to itself, is also close to the walk sequences. Blind walk is close to jogging and walking. The crawl sequence does not match any of the rest and this is clear from row 13 of the matrix. Thus, the results obtained using our method are reasonably close to what we would expect from a human observer.

Next, we consider the situation where we try to recognize activities when the input video sequences are from different viewpoints. This is the most interesting part of the method, as it demonstrates the strength of using 3D models for activity recognition. In our data set, we had three sequences where the motion is not parallel to the image plane: two for jogging in a circle and one for brooming in a circle. We considered a portion of these sequences where the person is not parallel to the camera. From each such video sequence, we computed the basis shapes. This basis shape is rotated, based on an estimate of its pose, and transformed to the canonical plane (i.e. parallel to the image plane). The basis shapes before and after rotation are shown in Fig. 4.8. This rotated basis shape is used to compute the similarity of this sequence with the others, exactly as described above. Rows 14–16 of the similarity

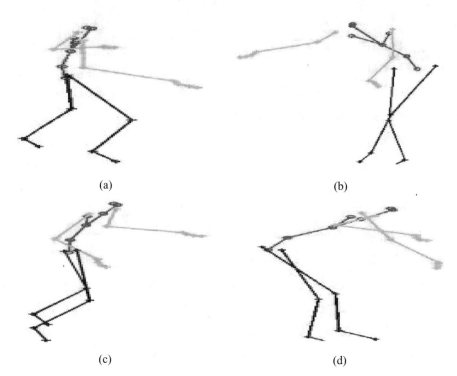

(a)

(b)

(c)

(d)

FIGURE 4.8: (a) and (b) The basis shapes for jogging and brooming when the viewing direction is different from the canonical one. (c) and (d) The rotated basis shapes.

matrix shows the recognition performance for this case. The jogging sequences are close to the jogging in the canonical plane (column 11), followed by walking along the canonical plane (columns 1–6). For the broom sequence, it is closest to the brooming in the canonical plane (columns 9 and 10). The sequences in Columns 17 and 18 are close to the other sitting and walking sequences, even though they are captured from different viewing directions.

4.3.3.2 Activity Recognition for Video Surveillance

In this section, we consider a very different kind of activity. A group of people get off an airplane and walk to the terminal. Also, there are other moving objects like vehicles, airport personnel, etc. The goal is to classify the activities of the different groups of objects (people vs. vehicles) and to identify an abnormal behavior (e.g. a passenger straying from the normal path), using the information available in the trajectories. The approach has a distinct training phase, followed by a testing phase.

Given F frames of a video sequence with moving points representing different activities, we can obtain the trajectories of all these points over the entire video sequence. An average trajectory for each of the activities can then be obtained. The trajectory defines the particular activity. For the case of people getting off an airplane, each person is represented by a point. An average trajectory over all the people represents the activity of people getting of the plane. If we have M different *training* video sequences with different instances of the same activity, we obtain many such example trajectories. Each of the example trajectories can be sampled uniformly to produce a set of P points, each represented as a pair of x and y coordinates, for each video sequence. Note that the number of rows in the matrix $\mathbf{W}$ in (4.11) depends on the number of training sequences, i.e. $F = M$.

During training, we compute the rotation matrix and the average shapes as explained above. For the mth video sequence, consider the rows $(2m - 1)$ and $2m$ of the matrix $\mathbf{W}$, and represent it by W_m. It represents the average trajectory of the activities in the mth training sequence. From (4.12), we see that $l_{m,i}$ can be computed by taking the inner product of W_m with $R_m S_i$, i.e.

$$l_{m,i} = <W_m, R_m S_i>$$
(4.16)

for each activity $i = 1, \ldots, N$ and for each training video sequence $m = 1, \ldots, M$. Thus for each activity i, we have M values of l_i. These multiple values of l_i represent a significant part of the range of values that can be taken by different instances of these activities. Since a fixed camera is looking at the same set of activities, the rotation matrices will not be very different between the different instances of the same activity. Hence, all the l_i for each activity cluster together and can be used for recognition.

During testing, we consider the trajectory of each object in the video sequence. The procedure described above can be reapplied to the set of tracked points in the sequence in order to obtain the configuration weights by projecting onto the rotated basis shapes, as in (4.16). The cluster to which the computed l_i belong can be used to identify the activity. The intuitive idea is that the set of weights learned from the training examples cover most of the possible ones for normal activities. Thus, if projections for the test activity lie within a cluster for one of the activities, then we claim to have recognized that particular activity. In practice, we set a threshold, $T < M$, for the number of projections that need to lie within a cluster for the activity to be recognized as such. By this method, the activity of each object is individually detected and verified in this 3D shape space. One of the advantages of this method is that it is computationally very inexpensive, since all that it does for classification and verification is to compute projections of tracked features onto basis shapes learned *a priori*.

We consider an airport surveillance situation, which has been described before. The trajectories of the main objects are obtained using a motion detection and tracking algorithm. In Fig. 4.9(a), we plot the average centered shapes (i.e. after the mean of every row of W is subtracted out) for the two major activities, passengers disembarking and the path of the luggage cart or fuel tank. The airport personnel

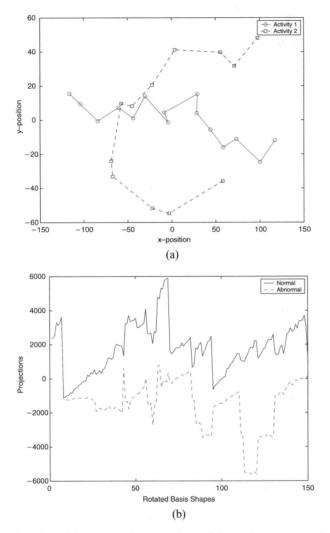

FIGURE 4.9: (a) Plot of the centered shapes formed from the average trajectories of the two activities. (b) Projections of the abnormal activity and a normal one on the rotated basis shapes for the first activity.

are identified *a priori* and their motion is neglected for the purposes of this analysis. It is clear from the plot that the shapes are very different, and successfully exploiting them can lead to a good classification algorithm for the various activities. Also, when an abnormal event occurs [Fig. 4.2(b)], the trajectory, as represented by the shape, is significantly deformed and can be identified.

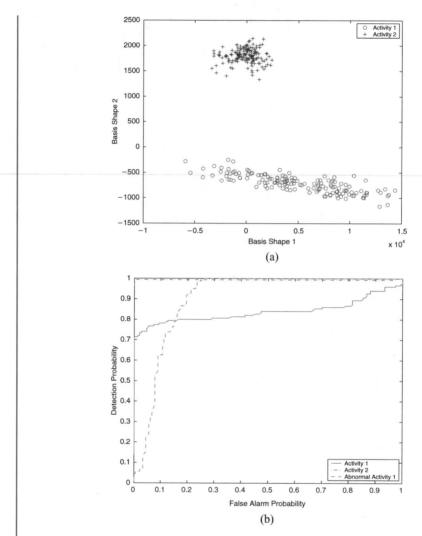

FIGURE 4.10: (a) Plot of the projections of the various instances of the two activities, as available in the training data, onto the rotated basis shapes. (b) ROC plots for classification of the two normal activities and the abnormal one.

The plot of the various values of $l_{m,1}$ and $l_{m,2}$ for all m, learned from the training sequences, is shown in Fig. 4.10(a), thus showing the clear demarcation between the two activities. In Fig. 4.11(a), we show the plots of the projections of the activity of passengers deplaning on the two sets of rotated basis shapes, learned during the training phase, i.e. $R_m S_1$ and $R_m S_2$, for $m = 1, \ldots, 150$. Another test

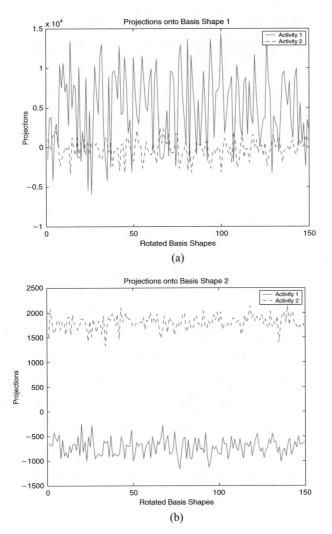

FIGURE 4.11: Projections of the two activities on the rotated basis shapes for the first one are shown in (a), while the projections on the rotated basis shapes for the second one are shown in (b).

case is the motion of the luggage cart. Its projections on the two sets of rotated basis shapes is shown in Fig. 4.11(b). The plots in Fig. 4.11 can be used to distinguish between the two activities, given just their motion trajectories by setting an appropriate threshold and declaring an activity to be either one or two, depending on the number of points on either side of the threshold. We can thus automatically verify whether each of the different tasks, like passengers boarding a plane or luggage loaded into the cargo hold and the cart departing, were completed successfully or not.

The next task is to determine any abnormalities. By this we mean the detection of the case shown in Fig. 4.2(b). Since the testing is done for each object at a time, the process can identify the concerned individual or object. As we do not have real video sequences of such behavior, we simulated it by pulling a passenger away from the normal path. Fig. 4.9(b) plots the projections for the abnormal activity and a normal one on the set of rotated basis shapes. The clear difference in the projections shows the difference in the two activities, which can help to identify the abnormal one.

The receiver operating characteristic (ROC) of the activity detection algorithm is shown in Fig. 4.10(b). The plots are obtained through simulations by varying the threshold of detection for the two normal activities, as well as the abnormal one. For classification between the two activities, a detection occurs when a test activity, say A, is recognized correctly from the projections onto the set of rotated basis shapes of A, while a false alarm is defined as the case when the projections onto the rotated basis shapes of A of the trajectory obtained from some other activity exceeds the detection threshold. For an abnormal activity, a detection occurs when it is correctly identified as abnormal, while a false alarm occurs when a normal activity is flagged as abnormal.

4.4   VIDEO SUMMARIZATION

We performed an experiment to summarize a 3 minute segment of video obtained for the airport surveillance example in the activity shape space using the subspace analysis method. The motion trajectories of all moving objects were considered. They included the passengers, a luggage cart, and an airport personnel (whose motion has not been modeled as part of the training procedure, but who can be seen at the bottom of Fig. 4.2(a)). The motion trajectory of each individual object was projected onto the set of rotated basis shapes $R_m S_i$, for $m = 1, \ldots, 150$, $i = 1, 2$

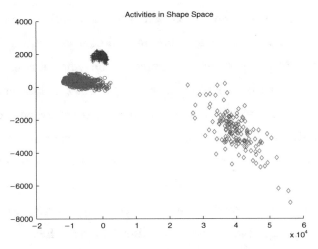

FIGURE 4.12: A video summarization example: projections of all the motion trajectories in a 3-min segment of the video sequence onto the basis shapes. The red cluster contains the projections of the passengers, the blue of the luggage cart, and magenta of the airport personnel whose motion was not modeled as part of the training examples.

learned from the training examples, as explained before. Fig. 4.12(b) shows the projections form three clusters, corresponding to the motion trajectories of 10 passengers, the luggage cart, and an airport personnel. These three clusters contain information about all the moving objects in the 3 minute segment of the video. The clusters can also be useful for identifying an abnormal activity, which does not lie in any of the clusters learned for the set of normal activities. Hence we see that it is possible to summarize the motion of all objects in the scene in the shape space.

4.5 DISCUSSION: SHAPE-BASED ACTIVITY MODELS

We now address two questions related to the approaches described in this chapter. What is the justification of modeling an activity by a nonrigid shape and its dynamics? Is it necessary to use 3D models for this process, as opposed to 2D appearance based approaches?

The use of shape models for activity is based on the empirical observation that many activities have an associated structure and a dynamical model. Consider, as an example, a dancer or figure skater, who is free to move her hands and feet any way she likes. However, this random movement does not constitute the activity of dancing. For humans to perceive and appreciate the dance, the different parts of the body have to move in a certain synchronized manner. In mathematical terms, this is equivalent to modeling the dance by the structure of the body of the dancer and its dynamics. Similar comments can be made for other activities performed by a single human, e.g. walking, jogging, sitting, etc. An analogous example exists in the domain of video surveillance. Consider the example of people getting off a plane and walking to the terminal, where there is no jet-bridge to constrain the path of the passengers. Every person after disembarking, is free to move as he/she likes. However, this does not constitute the activity of people getting off a plane and heading to the terminal. The activity here is comprised of people walking along a path that leads to the terminal. Again, we see that the activity is defined by a structure and the dynamics associated with the structure. Using a shape-dynamical model is a higher level abstraction of the individual trajectories and provides a method of analyzing all the points of interest together, thus modeling their interactions in a very elegant way.

The next question that can be raised is the following: Do we need to build 3D models of shape, which are often not easy to obtain, is order to perform activity classification accurately? We explained in Chapter 3 our experiments to understand the role of shape and dynamics in human activity inference. We will use the framework of Section 3.4 in order to understand the role of 3D models.

Consider the vector of points representing the activity in each frame to be $\alpha(t), t = 1, \ldots, F$. First we consider an autoregressive (AR) model on these points, i.e. $\alpha(t) = A\alpha(t-1) + w(t)$, w is a zero mean white Gaussian noise process and A is the transition matrix. If A_j and B_j (for $j = 1, 2, \ldots N$) represent the transition matrices for two sequences representing two activities, then the distance between models is defined as $D(A, B)$:

$$D(A, B) = \sum_{j=1}^{j=N} ||A_j - B_j||_{\mathrm{F}}, \tag{4.17}$$

where $||.||_F$ denotes the Frobenius norm. The model in the gallery that is closest to the model of the given probe is chosen as the identity of the person.

Next an ARMA model on the points is used. This linear dynamical model can be represented as

$$\alpha(t) = Cx(t) + w(t); w(t) \sim N(0, R) \tag{4.18}$$

$$x(t + 1) = Ax(t) + v(t); v(t) \sim N(0, Q). \tag{4.19}$$

Let the cross correlation between w and v be given by S. The parameters of the model are given by the transition matrix A and the state matrix C. We note that the choice of matrices A, C, R, Q, S is not unique. But, we also know (121) that we can transform this model to the "innovation representation." The model parameters in the innovation representation are unique. The model parameters are learned using the algorithm described in (121) and (155). The distance between two ARMA models, ($[A_1, C_1]$ and $[A_2, C_2]$) is computed using subspace angles (56) and as described in (30).

The plots of the similarity matrices for the activities in the MOCAP data using the AR and ARMA models are shown in Fig. 4.13. Note that the AR model uses purely dynamical information, while the ARMA model encodes 2D shape information also in the C matrix. Comparing these two figures leads to the following conclusion: A pure dynamical model (AR) has less discriminating power than an ARMA model. For example, all the walk sequences in Fig. 4.13(a) are not grouped together, as they should be. However, comparison of these two figures with Fig. 4.7 shows that the use of the 3D model increases the recognition performance even when there is no change in viewing direction. For example, the similarities between jogging and walking are clearer than when using an ARMA model. Also, crawling is clearly distinct from the rest in Fig. 4.7. Hence there is a clear advantage in using 3D models over 2D models for activity classification. In addition, 3D models allow view invariant recognition and multicamera use, as we have explained before.

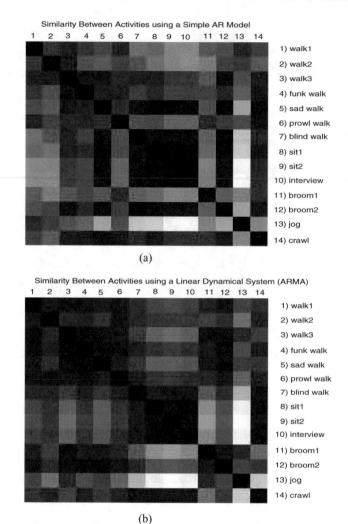

FIGURE 4.13: Plot of the similarity matrix for activity classification using (a) AR models, and (b) ARMA model.

4.6 CONCLUSION

Activity recognition from video is a very broad area of research and touches upon different aspects of computer vision. It is impossible to do justice to all the existing methods within a few pages. We have provided a broad review of existing techniques and described two of the methods that we have developed. Both of these approaches are based on a nonrigid shape model for each activity. One of the methods is appearance based, the other is 3D model based. Present work in this area is largely concentrated on the use of distributed networks of video cameras, which can provide interpretations from various viewpoints. However, that introduces it own problems of handling large amounts of data in an efficient manner. Also, issues of quality of data coming in from the sensors and the bandwidth capacity required for multiple sensors are not yet fully understood.

CHAPTER 5

Future Research Directions

From the previous chapters, it is clear that there still exist a number of issues that need to be resolved in identifying humans and their activities from video. Some of the most pressing problems are illuminations and pose variations. One of the approaches to deal with this problem is building a 3D model from video. This is however still a difficult problem under unconstrained situations. Quality of video data, efficiency of the reconstruction algorithm, and processing time stand in the way of use of 3D models in recognition. In this chapter, we outline two methods for 3D modeling. First we describe a method for face modeling using SfM, where the quality of the video data is taken into account. Next, we describe a method for modeling the human body and its motion using kinematic chains in a multicamera framework. Finally, we demonstrate how the use of different modalities, face and gait for instance, can lead to higher recognition performance.

5.1 THREE-DIMENSIONAL FACE MODELING USING SfM

Various researchers have addressed the issue of 3D face modeling. In (75), the authors used an extended Kalman filter to recover the 3D structure of a face, which was then used for tracking. A method for recovering nonrigid 3D shapes as a linear combination of a set of basis shapes was proposed in (21). A factorization based method for recovering nonrigid 3D structure and motion from video was presented in (19). In (133), the author proposes a method for self calibration in the presence of varying internal camera parameters and reconstructs metric 3D structure. Romdhani *et al.* have shown that it is possible to recover the shape and texture parameters of a 3D morphable model from a single image (138). They used their 3D model for identification of faces under different pose and illumination conditions. Our method (143; 145) is along the lines of (49) and (149), where the authors proposed solving the problem of 3D face modeling using a generic model. Their method of bundle-adjustment works by initializing the reconstruction algorithm with this generic model. The difficulty with this approach is that the algorithm often converges to a solution very near this initial value, resulting in a reconstruction that has the characteristics of the generic model, rather than that of the particular face in the video that needs to be modeled. This method may give very good results when the generic model has significant similarities with the particular face being reconstructed. However, if the features of the generic model are different from those of the face being reconstructed, the solution obtained using this approach may be unsatisfactory. We also estimate the quality of the input video and incorporate it into the reconstruction algorithm.

5.1.1   Error Estimation in 3D Reconstruction

Since the motion between adjacent frames in a video sequence of a face is usually small, we will adopt the optical flow framework for reconstructing the structure (118). It is assumed that the coordinate frame is attached rigidly to the camera with the origin at the center of perspective projection and the z-axis perpendicular to the image plane. The camera is moving with respect to the face being modeled (which is assumed rigid) with translational velocity $\mathbf{V} = [v_x, v_y, v_z]$ and rotational velocity $\Omega = [\omega_x, \omega_y, \omega_z]$ (this can be reversed by simply changing the sign of the velocity vector). Using the small-motion approximation to the perspective projection model for motion field analysis, and denoting by $p(x, y)$ and $q(x, y)$, the horizontal and vertical velocity fields of a point (x, y) in the image plane, we can write the equations relating the object motion and scene depth by (118)

$$p(x, y) = (x - fx_f)h(x, y) + \frac{1}{f}xy\omega_x - (f + \frac{1}{f}x^2)\omega_y + y\omega_z$$

$$q(x, y) = (y - fy_f)h(x, y) + (f + \frac{1}{f}y^2)\omega_x - \frac{1}{f}xy\omega_y - x\omega_z, \quad (5.1)$$

where f is the focal length of the camera, $(x_f, y_f) = (\frac{v_x}{v_z}, \frac{v_y}{v_z})$ is known as the *focus of expansion* (FOE), and $h(x, y) = \frac{v_z}{z(x,y)}$ is the scaled inverse scene depth. We will assume that the FOE is known over a few frames of the video sequence. Under the assumption that the motion between adjacent frames in a video is small, we compute the FOE from the first two or three frames and then keep it constant over the next few frames (157). For N corresponding points, using subscript i to represent the above defined quantities at the ith point, we define (similar to (157))

$$\mathbf{h} = (h_1, h_2, \ldots, h_N)^T_{N \times 1}$$

$$\mathbf{u} = (p_1, q_1, p_2, q_2, \ldots, p_N, q_N)^T_{2N \times 1}$$

$$\mathbf{r}_i = (x_i y_i, -(1 + x_i^2), y_i)^T_{3 \times 1}$$

$$\mathbf{s}_i = (1 + y_i^2, -x_i y_i, -x_i)^T_{3 \times 1}$$

$$\Omega = (w_x, w_y, w_z)^T_{3 \times 1}$$

$$\mathbf{Q} = \begin{bmatrix} r_1 & s_1 & r_2 & s_2 & \cdots & r_N & s_N \end{bmatrix}^T_{2N \times 3}$$

$$\mathbf{P} = \begin{bmatrix} x_1 - x_f & 0 & \cdots & 0 \\ y_1 - y_f & 0 & \cdots & 0 \\ 0 & x_2 - x_f & \cdots & 0 \\ 0 & y_2 - y_f & \cdots & 0 \\ \vdots & \vdots & \ddots & \vdots \\ 0 & 0 & \cdots & x_N - x_f \\ 0 & 0 & \cdots & y_N - y_f \end{bmatrix}_{2N \times N}$$

$$\mathbf{A} = [\mathbf{P} \quad \mathbf{Q}]_{2N \times (N+3)}$$

$$\mathbf{z} = \begin{bmatrix} \mathbf{h} \\ \Omega \end{bmatrix}_{(N+3) \times 1}. \quad (5.2)$$

Then (5.1) can be written as

$$\mathbf{Az} = \mathbf{u}. \tag{5.3}$$

Our aim is to compute $\mathbf{z}$ from $\mathbf{u}$ and to obtain a quantitative idea of the accuracy of the 3D reconstruction $\mathbf{z}$ as a function of the uncertainty in the motion estimates $\mathbf{u}$. Let us denote by $\mathbf{R_u}$ the covariance matrix of $\mathbf{u}$ and by C the cost function

$$C = \frac{1}{2}||\mathbf{Az} - \mathbf{u}||^2 = \frac{1}{2}\sum_{i=1}^{n=2N} C_i^2(u_i, \mathbf{z}). \tag{5.4}$$

In (143), using the implicit function theorem (180), we proved the following result.

Theorem 1 *Define*

$$\begin{aligned}
\mathbf{A}_{\bar{i}p} &= [0 \cdots 0 \, -(x_i - x_f) \, 0 \cdots 0 \, -x_i y_i \, (1 + x_{\bar{i}}^2) \, -y_i], \\
&= [-(x_i - x_f)\mathbf{I}_i(N)\,|\,-\mathbf{r}_i] = [A_{\bar{i}ph}|A_{\bar{i}pm}] \\
\mathbf{A}_{\bar{i}q} &= [0 \cdots 0 \, -(y_i - y_f) \, 0 \cdots 0 \, -(1 + y_{vi}^2) \, x_i y_i(N) \, x_i], \\
&= [-(y_i - y_f)\mathbf{I}_i(N)\,|\,-\mathbf{s}_i] = [A_{\bar{i}qh}|A_{\bar{i}qm}]
\end{aligned} \tag{5.5}$$

where $\bar{i} = \lceil i/2 \rceil$ is the ceiling of i ($\bar{i}$ will then represent the number of feature points N and $i = 1, \ldots, n = 2N$) and $\mathbf{I}_n(N)$ denotes a 1 in the nth position of the array of length N and zeros elsewhere. The subscript p in $\mathbf{A}_{\bar{i}p}$ and q in $\mathbf{A}_{\bar{i}q}$ denotes that the elements of the respective vectors are derived from the pth and qth components of the motion in (5.1). Then

$$\mathbf{R_z} = \mathbf{H}^{-1}\left(\sum_i \frac{\partial C_i^T}{\partial \mathbf{z}}\frac{\partial C_i}{\partial \mathbf{u}}\mathbf{R_u}\frac{\partial C_i^T}{\partial \mathbf{u}}\frac{\partial C_i}{\partial \mathbf{z}}\right)\mathbf{H}^{-T} \tag{5.6}$$

$$= \mathbf{H}^{-1}\left(\sum_{\bar{i}=1}^{N}\left(\mathbf{A}_{\bar{i}p}{}^T\mathbf{A}_{\bar{i}p}R_{u\bar{i}p} + \mathbf{A}_{\bar{i}q}{}^T\mathbf{A}_{\bar{i}q}R_{u\bar{i}q}\right)\right)\mathbf{H}^{-T}, \tag{5.7}$$

and

$$\mathbf{H} = \sum_{\bar{i}=1}^{N}\left(\mathbf{A}_{\bar{i}p}{}^T\mathbf{A}_{\bar{i}p} + \mathbf{A}_{\bar{i}q}{}^T\mathbf{A}_{\bar{i}q}\right), \tag{5.8}$$

where $\mathbf{R_u} = \mathrm{diag}\left[R_{u1p}, R_{u1q}, \ldots, R_{uNp}, R_{uNq}\right]$.

Because of the partitioning of $\mathbf{z}$ in (5.2), we can write

$$\mathbf{R_z} = \begin{bmatrix} \mathbf{R_h} & \mathbf{R_{hm}} \\ \mathbf{R_{hm}^T} & \mathbf{R_m} \end{bmatrix}. \tag{5.9}$$

We can then show that for N points and M frames, the average distortion in the reconstruction is

$$D_{\mathrm{avg}}(M, N) = \frac{1}{MN^2} \sum_{j=1}^{M} \mathrm{trace}(\mathbf{R_h^j}), \tag{5.10}$$

where the superscript in the index to the frame number. We will call (5.10) the multiframe SfM (MFSfM) rate-distortion function, hereafter referred to as the video rate-distortion (VRD) function. Given a particular tolerable level of distortion, the VRD specifies the minimum number of frames necessary to achieve that level. In (141; 144) we had proposed an alternative information theoretic criterion for evaluating the quality of a 3D reconstruction from video and had analyzed the comparative advantages and disadvantages. The above results do not require the standard assumptions of gaussianity of observations and is thus an extension of the error covariance results presented in (187).

5.1.2   SfM Algorithm for Face Reconstruction

Fig. 5.1 shows a block diagram schematic of the complete 3D face reconstruction framework using SfM. The input is a monocular video sequence. We choose an appropriate two-frame depth reconstruction strategy (157). The depth maps are aligned to a single frame of reference and the aligned depth maps are fused together using stochastic approximation.

Let $\mathbf{s}^i \in \mathbf{R}^3$ represent the structure,[1] computed for a particular point, from ith and $(i + 1)$st frame, $i = 1, \ldots, K$, where the total number of frames is $K + 1$.[2]

---

[1] In our description, subscripts will refer to feature points and superscripts will refer to frame numbers. Thus x_i^j refers to the variable x for the ith feature point in the jth frame.

[2] For notational simplicity, we use i and $(i + 1)$st frames to explain our algorithm. However, the method can be applied for any two frames provided the constraints of optical flow are not violated.

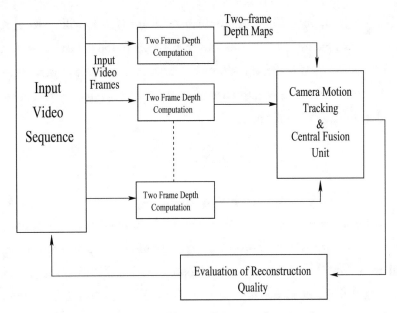

FIGURE 5.1: Block diagram of the 3D reconstruction framework.

Let the fused structure subestimate at the ith frame be denoted by $\mathbf{S}^i \in \mathbf{R}^3$. Let Ω^i and $\mathbf{V}^i$ represent the rotation and translation of the camera between the ith and $(i+1)$st frames. Note that the camera motion estimates are valid for all the points in the object in that frame. The 3×3 rotation matrix $\mathbf{P}^i$ describes the change of coordinates between times i and $i+1$, and is orthonormal with positive determinant. When the rotational velocity Ω is held constant between time samples, $\mathbf{P}$ is related to Ω by $\mathbf{P} = e^{\hat{\Omega}}$.[3] The fused subestimate $\mathbf{S}^i$ can now be transformed as $T^i(\mathbf{S}^i) = \mathbf{P}^i \mathbf{S}^i + \mathbf{V}^{i\,T}$. But in order to do this, we need to estimate the motion parameters $\mathbf{V}$ and Ω. Since we can determine only the direction of translational motion $(v_x/v_z, v_y/v_z)$, we will represent the motion components by the vector

---

[3] For any vector $\mathbf{a} = [a_1, a_2, a_3]$, there exists a unique skew-symmetric matrix

$$\hat{a} = \begin{bmatrix} 0 & -a_3 & a_2 \\ a_3 & 0 & -a_1 \\ -a_2 & a_1 & 0 \end{bmatrix}.$$ (5.11)

The operator $\hat{a}$ performs the vector product on $\mathbf{R}^3$: $\hat{a}X = \mathbf{a} \times X, \forall X \in \mathbf{R}^3$. With an abuse of notation, the same variable is used for the random variable and its realization.

$\mathbf{m} = \left[\frac{v_x}{v_z}, \frac{v_y}{v_z}, \omega_x, \omega_y, \omega_z\right]$. Thus, the problems at stage $(i + 1)$ will be to (i) reliably track the motion parameters obtained from the two-frame solutions and (ii) fuse $\mathbf{s}^{i+1}$ and $T^i(\mathbf{S}^i)$. If $\{l^i\}$ is the transformed sequence of inverse depth values with respect to a common frame of reference, then the optimal value of the depth at the point under consideration is obtained as

$$u^* = \arg \min_u \operatorname{median}_i \left(w_l^i (l^i - u)^2 \right), \qquad (5.12)$$

where $w_l^i = (\mathrm{R_h^i}(l))^{-1}$, with $\mathrm{R_h^i}(l)$ representing the covariance of l^i [which can be obtained from (5.9)]. However, since we will be using a recursive strategy, it is not necessary to align all the depth maps to a common frame of reference *a priori*. We will use a Robbins–Monro stochastic approximation (RMSA) (137) algorithm (refer to (143) for details) where it is enough to align the fused subestimate and the two-frame depth for each pair of frames and proceed as more images become available.

For each feature point, we compute $X^i(u) = w_l^i(l^i - u)^2, u \in \mathcal{U}$. At each step of the RM recursion, the fused inverse depth, $\hat{\theta}^{k+1}$, is updated according to (143)

$$\hat{\theta}^{k+1} = \hat{T}^k(\hat{\theta}^k) - a^k(p^k(\hat{\theta}^k) - 0.5), \qquad (5.13)$$

where a^k is determined by a convergence condition , $p^k(\hat{\theta}^k) = \mathbf{I}_{[X^k \leq \hat{T}^k(\hat{\theta}^k)]}$, $\mathbf{I}$ represents the indicator function, and $\hat{T}^k$ is the estimate of the camera motion. When $k = K$, we obtain the fused inverse depth $\hat{\theta}^{K+1}$, from which we can get the fused depth value $\mathbf{S}^{K+1}$. The camera motion $\hat{T}$ is estimated using a tracking algorithm as described in (141).

The Reconstruction Algorithm: Assume that we have the fused 3D structure $\mathbf{S}^i$ obtained from i frames and the two-frame depth map $\mathbf{s}^{i+1}$ computed from the ith and $(i + 1)$st frames. Fig. 5.2 shows a block diagram of the multiframe fusion algorithm. The main steps of the algorithm are as follows:

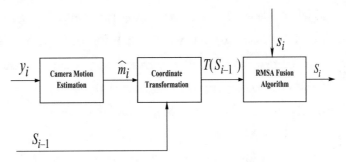

FIGURE 5.2: Block diagram of the multiframe fusion algorithm.

Track Estimate the camera motion using the camera motion tracking algorithm (141).

Transform Transform the previous model $\mathbf{S}^i$ to the new reference frame.

Update Update the transformed model using $\mathbf{s}^{i+1}$ to obtain $\mathbf{S}^{i+1}$ from (5.13).

Evaluate Reconstruction Compute a performance measure for the fused reconstruction from (5.10).

Iterate Decide whether to stop on the basis of the performance measure. If not, set $i = i + 1$ and go back to **Track**.

5.1.3   Incorporating the Generic Face Model

The Optimization Function: Both the generic model and the 3D estimate have a triangular mesh representation with N vertices and the depth at each of these vertices is known. (We explain how this can be obtained in a later section.) Let $\{d_{g_i}, i = 1, \dots, N\}$ be the set of depth values of the generic mesh for each of the N vertices of the triangles of the mesh. Let $\{d_{s_i}, i = 1, \dots, N\}$ be the corresponding depth values from the SfM estimate. We wish to obtain a set of values $\{f_i, i = 1, \dots, N\}$ that are a smoothed version of the SfM model, after correcting the errors on the basis of the generic mesh.

Since we want to retain the specific features of the face we are trying to model, our error correction strategy works by comparing local regions in the two models and smoothing those parts of the SfM estimate where the *trend* of the depth values is significantly different from that in the generic model, e.g. a sudden peak on the forehead will be detected as an outlier after the comparison and smoothed. This is where we differ from previous work (49; 149), since we do not intend to fuse the depth in the two models but to correct errors based on local geometric trends. Toward this goal, we introduce a line process on the depth values. The line process indicates the borders where the depth values have sudden changes, and is calculated on the basis of the generic mesh, since it is free from errors. For each of the N vertices, we assign a binary number indicating whether or not it is part of the line process. This concept of the line process is borrowed from the seminal work of Geman and Geman (54) on stochastic relaxation algorithms in image restoration.

The optimization function we propose is

$$E(f_i, l_i) = \sum_{i=1}^{N} (f_i - d_{s_i})^2 + (1 - \mu) \sum_{i=1}^{N} (f_i - d_{g_i})^2$$

$$+ \mu \sum_{i=1}^{N} (1 - l_i) \sum_{j \in \mathcal{N}_i} (f_i - f_j)^2 \mathbf{1}_{d_s \neq d_g}, \qquad (5.14)$$

where $l_i = 1$ if the ith vertex is part of a line process and μ is a combining factor, which controls the extent of the smoothing. $\mathcal{N}_i$ is the set of vertices, which are neighbors of the ith vertex. $\mathbf{1}_{d_s \neq d_g}$ represents the indicator function, which is 1 if $d_s \neq d_g$, else 0. In order to understand the importance of (5.14), consider the third term. When $l_i = 1$, the ith vertex is part of a line process and should not be smoothed on the basis of the values in $\mathcal{N}_i$; hence, this term is switched off. Any errors in the value of this particular vertex will be corrected on the basis of the first two terms, which control how close the final smoothed mesh will be to the generic one and the SfM estimate. When $l_i = 0$, indicating that the ith vertex is not part of a line process, its final value in the smoothed mesh is determined by the neighbors as well as its corresponding values in the generic model and SfM estimate. The importance of each of these terms is controlled by the factor $0 < \mu < 1$. Note that μ is a function of d_s and d_g, and is computed by comparing the line process obtained from d_s with that pre-computed from d_g. In the case (largely academic) where $d_s = d_g$, the smoothed mesh can be either d_s or d_g and this is taken care of in the indicator function in the third term in (5.14). The line process l_i has a value of 1 or 0, depending on whether there is a sudden change in the depth in the generic model at the particular vertex i. Obtaining such changes relies on derivative computations, which are known to be noise prone. Hence, in our optimization scheme, we allow the line process to be perturbed slightly around its nominal value computed from the generic model.

We use the technique of simulated annealing built upon the Markov Chain Monte Carlo (MCMC) framework (38). MCMC is a natural method for solving energy function minimization problems (29). The MCMC optimizer is essentially a Monte Carlo integration procedure in which the random samples are produced by evolving a Markov chain. Let $T_1 > T_2 > \cdots > T_k > \cdots$ be a sequence of monotone decreasing temperatures in which T_1 is reasonably large and $\lim_{T_k \to \infty} = 0$. At each such T_k, we run N_k iterations of a Metropolis-Hastings (M-H) sampler (38) with the target distribution represented as $\pi_k(f, l) \propto \exp\{-E(f, l)/T_k\}$. As k increases, π_k puts more and more of its probability mass (converging to 1) in the vicinity of the global maximum of E. Since minimizing $E(f, l)$ is equivalent to maximizing $\pi(f, l)$, we will almost surely be in the vicinity of the global optimum if the number of iterations N_k of the M-H sampler is sufficiently large. The steps of the algorithm are as follows:

- Initialize at an arbitrary configuration f_0, l_0 and initial temperature level T_1. Set $k = 1$.

- For each k, run N_k steps of MCMC iterations with $\pi_k(f, l)$ as the target distribution. Consider the following update strategy. For the line process, consider all the vertices (say $L < N$) for which the nominal value, $l_{i,\text{nominal}} = 1$, and their individual neighborhood sets, $N_1, \ldots, N_L$. For each $l_i = 1$, consider the neighborhood set among $N_1, \ldots, N_L$ that it lies in, randomly choose a vertex in this neighborhood set whose value is not already set to 1, and switch the values of l_i and this chosen vertex. Starting from $l_i = l_{i,\text{nominal}}$, this process ensures that the values of l_i do not move too

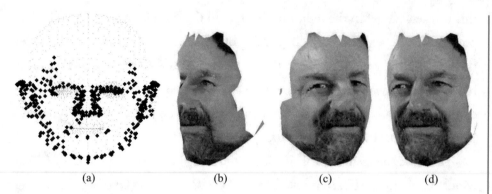

(a)	(b)	(c)	(d)

FIGURE 5.3: (a) The vertices which form part of the line processes indicating a change in depth values are indicated with ×. (b)–(d) Different views of the 3D model after texture mapping.

far from the nominal values. In fact, only the vertices lying in the neighborhood sets $N_1, \ldots, N_L$ can take a value of $l_i = 1$. Next, randomly determine a new value of f, using a suitable transition function (38). With the new values, $f_{\text{new}}, l_{\text{new}}$ of f, l, compute $\delta = E(f_{\text{new}}, l_{\text{new}}) - E(f, l)$. If $\delta < 0$, i.e. the energy decreases with this new configuration, accept $f_{\text{new}}, l_{\text{new}}$; else, accept with a probability ρ. Pass the final configuration of f, l to the next iteration.

- Increase k to $k + 1$.

The complete 3D reconstruction paradigm is composed of a sequential application of the two algorithms (3D reconstruction algorithm and the generic mesh algorithm) we have described in Sections 5.1.1 and 5.1.3. Some examples of 3D reconstruction are shown in Fig. 5.3. Details of the algorithm are available in (145).

5.2 MULTI-VIEW KINEMATIC CHAIN MODELS FOR HUMAN MOTION RECOVERY

There are several models for human body shape modelling from stick figures and ellipsoids to more complicated models that are deformable (78). Modelling the human body as rigid parts linked in a kinematic structure is a simple yet accurate for tracking purposes. Optical flow can be exploited to provide dense information and obtain robust estimates of the motion parameters. Several papers (22; 153; 185; 186) have employed this set of motion parameters. However, (22) uses orthographic projection, while (153) uses a Bayesian formulation combined with a particle filtering approach to determine the motion parameters. Yamamoto *et al.,* (186) use a different set of motion parameters to perform tracking. They use multiple views and perspective projection in their model, but have a larger parameter set and make some approximations in their formulation. We explain our approach to modeling human motion using a kinematic chain, which does not make some of the approximations in the above methods (163). Moreover, we propose an iterative algorithm by means of which we are able to arrive at an exact solution.

We model the movement of human beings using kinematic chains with the root of the kinematic tree being the torso (base body). Fig. 5.4 illustrates the model. Each body part forearm, upper arm, torso, head, etc., is modelled as a rigid body, whose shape is known and is expressed in an object reference frame. Usually

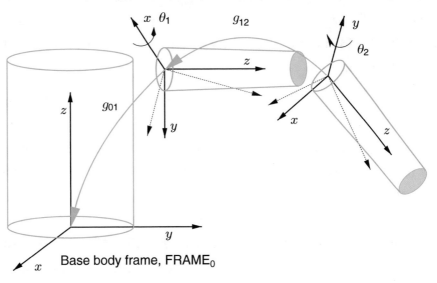

FIGURE 5.4: Kinematic chain schematic.

the limb is oriented along the z-axis of the object frame and the x or y axes are oriented so as to coincide with the axis around which the limb naturally rotates, if such an axis exists. The origins of the object coordinate system coincides with the joint. The object rotates about this point. The transformation from one object frame to the adjacent object frame can therefore be represented as a constant rigid transformation and a rotation that varies with time. At any given time we know the positions and orientations of the objects in the kinematic chain. We show that the instantaneous 3D velocity vector of each point in the base body coordinate system is a linear function of the vector comprising of V, the base body motion parameters, and $\dot{\theta}$, the vector angular velocities of each degree of freedom in the kinematic chain. In a computer vision system, however, we are able to measure only the 2D image velocities at each pixel for different cameras. We show that, under perspective projection, the 2D velocities are still linear in the 3D motion parameters and we can accurately compute this linear relation. This enables us to accurately obtain the 3D motion parameters of the kinematic system, given only the 2D pixel velocities obtained from multiple cameras. In a practical imaging system we cannot obtain the instantaneous velocities ($\dot{x}$) but can only obtain position differences (Δx). Errors are introduced at this stage, due to the approximation of $\dot{x}$ by Δx. In (163), we outlined an iterative algorithm by means of which we can reduce the error introduced by this approximation.

5.2.1   Obtaining the 3D Velocity

We define the base body reference frame, $FRAME_0$, that is attached to the base body (and moves with the base body) and a spatial reference frame, $FRAME_a$, that is static and coincides with $FRAME_0$ at time t_0. We would like to esti-mate the motion parameters at time t_0 (see Fig. 5.5). If we need to estimate the motion parameters at another time instant then $FRAME_a$ is appropriately redefined. We can view the motion of the base body as the motion of $FRAME_0$. We consider a single kinematic chain of J body parts connected to the base

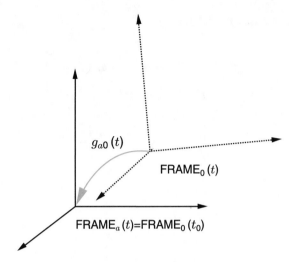

FIGURE 5.5: Motion of base body at time t.

body. (The equations that we derive can be easily extended to multiple such chains.) Each body part is indexed by a number i and attached to a coordinate frame $FRAME_i$ such that body i is connected to body $i-1$ and $i+1$, for $i = 1, \ldots, J-1$. The transformation from $FRAME_j$ to $FRAME_i$ is given by $g_{ij}(t)$. Consider a point on body part i given by $\boldsymbol{q}_i^{(i)}$. We have the following equations:

$$\boldsymbol{q}_a^{(i)}(t) = g_{a0}(t)\boldsymbol{q}_0^{(i)}(t)$$
$$\dot{\boldsymbol{q}}_a^{(i)}(t) = \dot{g}_{a0}(t)\boldsymbol{q}_0^{(i)}(t) + g_{a0}(t)\dot{\boldsymbol{q}}_0^{(i)}(t). \tag{5.15}$$

The instantaneous velocity of this point, as given in (5.15), has two components: one due to the motion of the base body itself (the first term on the RHS) and the other due to the motion of the kinematic chain (the second term). Since we model each of these motions differently, we deal with them separately.

5.2.1.1 Velocity Due to Motion of Base Body

We consider only the velocity component due to base body motion in this section. The motion of the base body is given by a rotation and a translation. From (5.15), dropping the point index i, we have

$$\hat{V}^s_{a0}\boldsymbol{q}_a = \begin{bmatrix} 0 & -\omega_3 & \omega_2 & v_1 \\ \omega_3 & 0 & -\omega_1 & v_2 \\ -\omega_2 & \omega_1 & 0 & v_3 \\ 0 & 0 & 0 & 0 \end{bmatrix} \begin{bmatrix} X_a \\ Y_a \\ Z_a \\ 1 \end{bmatrix}$$

$$= \begin{bmatrix} 1 & 0 & 0 & 0 & Z_a & -Y_a \\ 0 & 1 & 0 & -Z_a & 0 & X_a \\ 0 & 0 & 1 & Y_a & -X_a & 0 \\ 0 & 0 & 0 & 0 & 0 & 0 \end{bmatrix} V^s_{a0}$$

$$= \mathsf{W}(\boldsymbol{q}_a)V^s_{a0}, \tag{5.16}$$

where $\boldsymbol{q}_a(t) = [X_a(t), Y_a(t), Z_a(t), 1]^\mathsf{T}$ and $\hat{V}^s_{a0}$ is called the spatial velocity. It describes the translational velocity, $\boldsymbol{v}^s_{a0} = [v_1, v_2, v_3]^\mathsf{T}$, and the rotational velocity, $\omega^s_{a0} = [\omega_1, \omega_2, \omega_3]^\mathsf{T}$. For simplicity, we drop the dependence on time. Thus we see that the vector V^s_{a0} describes the motion of the base body and the instantaneous velocity of a point $\boldsymbol{q}_a$ is given by $\dot{\boldsymbol{q}}_a = \hat{V}^s_{a0}\boldsymbol{q}_a = \mathsf{W}(\boldsymbol{q}_a)V^s_{a0}$.

5.2.1.2 Velocity Due to Motion of Kinematic Chain

In a kinematic chain we have parts of the body connected by a kinematic chain. We parameterize the orientation between two connected components that possess a single degree of freedom in terms of the angle of rotation around the axis of the object coordinate frame, θ. We can deal with joints that have multiple degrees of freedom by considering them as multiple joints with single degree of freedom. Hereafter, we assume, without loss of generality, that all joints possess a single degree of freedom. This simplifies the form of the equations that follow.

Consider a point $\boldsymbol{q}$ represented by $\boldsymbol{q}_i$ in FRAME$_i$. Then the relation between $\boldsymbol{q}_0$ and $\boldsymbol{q}_i$ is given by $\boldsymbol{q}_0 = g_{0i}(t)\boldsymbol{q}_i$. The transformation of a point in FRAME$_0$ and FRAME$_i$ is represented by $g_{0i}(t)$. Then the instantaneous velocity of the point $\boldsymbol{q}_0$

is given by

$$\dot{\boldsymbol{q}}_0(t) = \hat{V}^s_{0i}\boldsymbol{q}_0(t), \tag{5.17}$$

where $\hat{V}^s_{0i}$ is the spatial velocity and is equal to $\dot{g}_{0i}(t)g^{-1}_{0i}(t)$.

In the case of kinematic chains, $g_{(i-1)i}(t)$ has the following form, where $\bar{g}_{(i-1)i} = g_{(i-1)i}(0)$:

$$g_{(i-1)i}(t) = \bar{g}_{(i-1)i}\, e^{\hat{\xi}_i\theta_i(t)}$$

$$\dot{g}_{(i-1)i}(t) = \hat{\xi}_i\dot{\theta}_i(t)\bar{g}_{(i-1)i}\, e^{\hat{\xi}_i\theta_i(t)}$$

$$\hat{V}^s_{(i-1)i}(t) = \dot{g}_{(i-1)i}(t)g^{-1}_{(i-1)i}(t)$$

$$= \bar{g}_{(i-1)i}(t)\hat{\xi}_i\dot{\theta}_i(t)\bar{g}^{-1}_{(i-1)i}(t).$$

Consider a kinematic chain with K links, where the motion of the kth link is represented by θ_k. The reference frame (indexed by subscript 0) is attached to the base body, and FRAME_k is attached to the kth body part, for $k \in \{1, 2, \ldots, K\}$. Note that if a joint has multiple degrees of freedom, then the number of body parts $(= J)$ is less than the number of frames $(= K)$. This does not alter the mathematical formulation in any way. We have $g_{0i} = g_{01}g_{12}\ldots g_{(i-1)i}$ and thus the instantaneous velocity, $\dot{\boldsymbol{q}}_0 = \hat{V}^s_{0i}\boldsymbol{q}_0$, where

$$\hat{V}^s_{0i} = \dot{g}_{0i}g^{-1}_{0i}$$

$$= \big(\dot{g}_{01}g_{12}\ldots g_{(i-1)i} + g_{01}\dot{g}_{12}\ldots g_{(i-1)i} + \cdots$$

$$\quad + g_{01}g_{12}\ldots \dot{g}_{(i-1)i}\big)\big(g_{01}g_{12}\ldots g_{(i-1)i}\big)^{-1}$$

$$= \dot{g}_{01}g^{-1}_{01} + g_{01}\big(\dot{g}_{12}g^{-1}_{12}\big)g^{-1}_{01} + \cdots$$

$$\quad + g_{0(i-1)}\big(\dot{g}_{(i-1)i}g^{-1}_{(i-1)i}\big)g^{-1}_{0(i-1)}$$

$$= \hat{V}^s_{01} + g_{01}\hat{V}^s_{12}g^{-1}_{01} + g_{02}\hat{V}^s_{23}g^{-1}_{02} + \cdots$$

$$\quad + g_{0(i-1)}\hat{V}^s_{(i-1)i}g^{-1}_{0(i-1)}.$$

We note that $q_0 = [X_0, Y_0, Z_0, 1]^\mathsf{T}$. We express the instantaneous velocity in terms of the kinematic chain motion parameters as

$$\begin{aligned}
\dot{q}_0 &= \hat{V}^s_{0i} q_0 \\
&= \hat{V}^s_{01} q_0 + \sum_{i=1}^{K-1} g_{0i} \hat{V}^s_{i(i+1)} g_{0i}^{-1} q_0 \\
&= \begin{bmatrix} \hat{\xi}'_1 & \hat{\xi}'_2 & \cdots & \hat{\xi}'_i & 0 & \cdots & 0 \end{bmatrix} \dot{\theta} \\
&= \mathsf{E}(q_0, \theta) \dot{\theta},
\end{aligned} \tag{5.18}$$

where $\dot{\theta} = \begin{bmatrix} \dot{\theta}_1 & \dot{\theta}_2 & \cdots & \dot{\theta}_K \end{bmatrix}$ and

$$\hat{\xi}'_j = \begin{cases} g_{0(j-1)} \bar{g}_{(j-1)j} \hat{\xi}_j \bar{g}^{-1}_{(j-1)j} g^{-1}_{0(j-1)} q_0 & j \leq i \\ 0 & j > i. \end{cases}$$

5.2.1.3 Combined 3D Velocity

We see that the complete 3D motion parameters are given by vectors V_{a0} and $\dot{\theta}$. Combining equations (5.15), (5.16), and (5.18), we get a linear relationship

$$\begin{aligned}
\dot{q}_a(t) &= \dot{g}_{a0}(t) q_0(t) + g_{a0}(t) \dot{q}_0(t) \\
&= \mathsf{W}(q_a) V_{a0} + g_{a0}(t) \mathsf{E}(g_{a0}^{-1} q_a, \theta) \dot{\theta} \\
&= \mathsf{M}(q_a, \theta) \varphi,
\end{aligned} \tag{5.19}$$

where $\mathsf{M}(q_a, \theta) = \begin{bmatrix} \mathsf{W}(q_a) & g_{a0}(t) \mathsf{E}(g_{a0}^{-1} q_a, \theta) \end{bmatrix}$ and $\varphi = \begin{bmatrix} V_{a0} \\ \dot{\theta} \end{bmatrix}$.

5.2.2 Multiple Camera Equations

In this section, we show that the instantaneous image velocity of a pixel, under perspective projection, is also linear in the motion parameters. Assume there are C cameras, numbered from 0 through $C - 1$. We use the superscript c to denote the camera index. $\mathsf{P}^{ca} = [p_1^{ca}, p_2^{ca}, p_3^{ca}]^\mathsf{T}$ maps a point in FRAME_a, $q_a = [X_a, Y_a, Z_a, 1]^\mathsf{T}$

to the homogeneous image coordinates, $\tilde{q}_c = [\tilde{x}_c, \tilde{y}_c, \tilde{z}_c]^T$, of the cth camera. We then have

$$\tilde{q}_c = \mathsf{P}^{ca} q_a.$$

The inhomogeneous image coordinates are given by

$$\begin{bmatrix} x_c \\ y_c \end{bmatrix} = \frac{1}{p_3^{ca\,T} q_a} \begin{bmatrix} p_1^{ca\,T} \\ p_2^{ca\,T} \end{bmatrix} q_a.$$

The image coordinate velocities can be derived as follows:

$$u_c = \begin{bmatrix} \dot{x}_c \\ \dot{y}_c \end{bmatrix}$$

$$= \frac{1}{p_3^{ca\,T} q_a} \begin{bmatrix} p_1^{ca\,T} \\ p_2^{ca\,T} \end{bmatrix} \dot{q}_a - \frac{p_3^{ca\,T} \dot{q}_a}{(p_3^{ca\,T} q_a)^2} \begin{bmatrix} p_1^{ca\,T} \\ p_2^{ca\,T} \end{bmatrix} q_a$$

$$= \frac{1}{\tilde{z}_c} \begin{bmatrix} p_1^{ca\,T} - (\tilde{x}_c/\tilde{z}_c) p_3^{ca\,T} \\ p_2^{ca\,T} - (\tilde{y}_c/\tilde{z}_c) p_3^{ca\,T} \end{bmatrix} \dot{q}_a$$

$$= \mathsf{C}^c(q_a, \mathsf{P}^{ca}) \dot{q}_a. \tag{5.20}$$

Combining (5.19) and (5.20), we get the following comprehensive equation:

$$u_c = \mathsf{C}^c(q_a, \mathsf{P}^{ca}) \mathsf{M}(q_a, \theta) \varphi. \tag{5.21}$$

We see that the number of unknowns is fixed, irrespective of the number of points and the number of cameras, and is equal to $6 + K$, where K is the number of degrees of freedom of the kinematic chain. If we can use $N(c)$ points from the cth camera, then we have a total of $2 \sum_{c=1}^{C} N(c)$ equations. Therefore we get two equation per pixel, q_a, per camera, c. Note that we need pixels on the kth body part to estimate $\theta_1, \ldots, \theta_k$. There is also no need to solve the correspondence problem across different cameras, as the only information required is the body part to which the pixel belongs.

5.3   FUSION OF FACE AND GAIT

Since performance is often not satisfactory, it is useful to have a number of independent cues that can then be fused together, either at the data level or the decision level, in order to achieve better performance (67; 86). We present here some basic experiments we have completed in fusion of face and gait.

In Section 2.3 a probabilistic approach to face recognition from video was described. The results of this method on the NIST database were presented. In Section 3.3, a view invariant approach to gait recognition was described and the results on the same database were presented. We will now show that combining these two results does improve overall performance of the recognition system.

The NIST database consists of 30 people walking along an inverted Σ-shaped walking pattern as shown in Fig. 3.7(b). The segment A is used as the gallery for gait recognition while the segment B (which is at an angle $33°$ to the horizontal part) is used as a probe. The gait recognition result (obtained by the procedure explained in Section 3.3) is shown in Figs. 5.6(a) and 5.6(d). The last part of the sequence where the person presents a front view to the camera (segment C) was used as the probe for still-to-video face recognition using (200). The gallery consisting of static faces for the 30 subjects. The results for face recognition are shown in Figs. 5.6(b) and 5.6(e).

In order to combine the scores from the face and gait classifiers directly, it is necessary to make them comparable. We used the exponential transformation for converting the scores obtained from the gait recognition, i.e., given that the match score for a probe X from the gallery gaits is given by $S_{X1}, \ldots, S_{XN}$ we obtain the transformed scores $\exp(-S_{X1}), \ldots, \exp(-S_{XN})$. Finally we normalize the transformed scores to sum to unity. We also tried logistic and logarithmic score transformation methods. The results obtained using these were comparable to the exponential case.

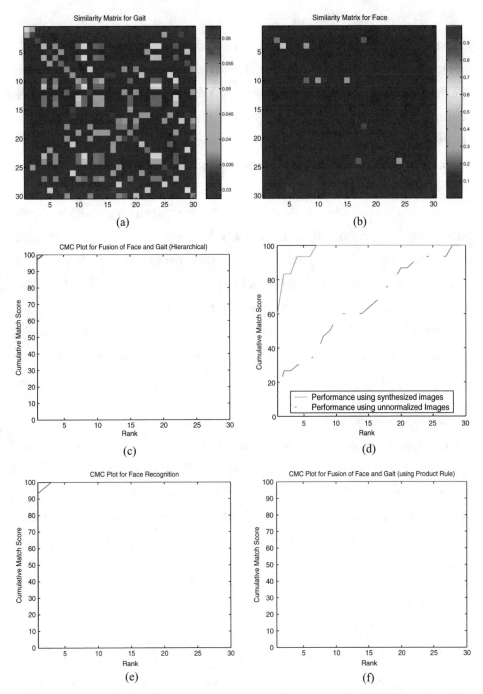

FIGURE 5.6: Similarity matrices for (a) gait recognition and (b)face recognition; CMC characteristics for (d) gait and (e) face; CMC curves for (c) hierarchical and (f) holistic fusion.

5.3.1 Hierarchical Fusion

Given the similarity matrix for the gait recognition algorithm, we plot the histograms of the diagonal and nondiagonal terms of the normalized similarity matrix. We note that these distributions have some overlap. This suggests that a threshold could be determined from the histogram and only individuals whose score is higher than this threshold need be passed to the face recognition algorithm. Although it is tempting to choose this threshold as high as possible, it should be noted that due to overlap in the two histograms, choosing a very high value may lead to the true person not being in the set of individuals passed to the face recognition algorithm. For the NIST database we chose a threshold of 0.035. This results in passing approximately the top six matches from the gait recognition unit to the face recognition algorithm. The CMC plot for the resulting hierarchical fusion is shown in Fig. 5.6(c). Note that the top match performance has gone up to 97% from 93% for this case. The more important gain however is in terms of the number of computations required. This number drops to one-fifth of its previous value. This demonstrates the value of gait as a filter.

5.3.2 Holistic Fusion

If the requirement from the fusion is that of accuracy as against computational speed, alternate fusion strategies can be employed. Assuming that gait and face can be considered to be independent cues, a simple way of combining the scores is to use the SUM or PRODUCT rule (86). Both the strategies were tried. The CMC curve for either case is as shown in Fig. 5.6(f). In both cases the recognition rate is 100%.

5.4   DEVELOPING AN ONTOLOGY OF ACTIVITY MODELING

Activities usually involve complex interactions between people and objects and the total range of possible human actions is very large. Hence, it is necessary to develop a scheme whereby various activites can be described according to certain general principles. A semantic representation of different events and their mutual relationships can provide such a framework. We provide a breif description of existing work in this area and the challenges involved in developing an accurate ontology of events.

There are certain common terminologies in semantic representations: dictionary, thesaurus, taxonomy, and ontology. Each of them have different strengths and weaknesses and so varying applications. Dictionary is simply a word list without any relation between words. In thesaurus, the words may only have equivalence relation between each other. Taxonomy is categorization of concepts, in which each component is placed on a hierarchical tree. In ontology, many distinct relations can be defined to represent the relations between concepts in the real world effectively. For instance, a car can have a "driven-by" relationship with human beings. At the same time the same car can have a "built-by" relationship with a company, which is operated by human beings.

Gruber defines five important criteria for ontology design as clarity, coherence, extendibility, minimal encoding bias, and minimal ontological commitment (60). Jones *et al.* define four different metrics for ontology evaluation, namely syntactic quality, semantic quality, pragmatic quality, and social quality. Each metric has individual attributes. Consistency and clarity are included as attributes of semantic quality. Pragmatic quality has attributes for amount, accuracy of information and its relevance for a given task. After scoring for individual definitions they calculate the overall quality by a weighted average (with weights depending on application) of these metrics (24).

In video surveillance, metadata is essential for formalizing a way to detect suspicious activities and providing a unifying framework for different methods in activity recognition (120). As detection of subevents is necessary to detect a composite event, and there are complex relations within those subevents with different constraints, ontology is the tool to be used if we want to formalize event detection. These constraints are spatiotemporal. For instance if someone is tailgating, his other entrance should follow an entry of another person authorized to enter the area (temporal constraint), and he or she should be hiding from the person entering before (spatial constraint). An initial ontology has been produced for six domains of video surveillance: perimeter and internal security, railroad crossing surveillance, visual bank monitoring, visual metro monitoring, store security, and airport-tarmac security in ARDA sponsored workshop "Challenge Project on Video Event Taxonomy" held in the summer and fall of 2003 La Jolla.

Yet the complexity of the ontology is dependent on the context of video surveillance. For instance, if we are think of a banking scenario, we see that all suspicious events can be characterized by unauthorized access to safe, somebody lying down for some reason (death, attack, lying down for protection, etc.). Hence a simple ontology will be enough for characterization of suspicious activities in this context. However, ontology is more complex for an airport tarmac security scenario, as the stakes are higher and any activity that does not follow the common procedures should be detected as a suspicious activity. An ontology that is capable of describing large classes of activities needs to consider all these issues.

C H A P T E R   6

Conclusions

Machine perception of humans and their actions from video is a very interesting problem from the perspective of a computer vision scientist. It spans a number of different areas, from low-level image processing tasks, to mid-level issues like depth, motion, and illumination estimation, to high-level tasks involving logical reasoning. We have touched upon a few of the approaches that have been adopted for the problems of face, gait, and activity recognition. We have described some existing methods, explained the unresolved issues, and outlined some of the approaches that we have developed. However, there are many issues that we did not address and which are becoming important in the research community. For example, the use of sophisticated cameras like omnidirectional cameras was not explored. Also, the use of sensor networks using different kinds of sensors was not touched upon. Video cameras, along with motion and weight sensors, are being employed in artistic performances in order to understand and model various human movements. This has spawned a new area called "digital arts." Research in these areas is still maturing, but they carry a significant potential for solving the problems in the future. If the goal of computer vision is to enable computers to see and understand the world, the perception of humans and their activities is one of most important tasks that will have to be resolved.

Face recognition is probably one of the topics in computer vision that has been successfully employed in certain real-life situations (though we still need to go a long way). Gait and activity modeling have very interesting implications for medicine, communications, virtual reality, and surveillance. The problems that confront researchers in all these areas are similar—variations of pose and illumination, occlusion, changes in clothing and disguise, and the effects of time. Automatic recognition of people and their activities has very important social implications, since it is related to the the extremely sensitive topic of civil liberties. This is an issue that society needs to address and come up with a balanced solution that is able to meet its various needs.

References

[1]   AFGR: *Proceedings of* IEEE *International Conference on Automatic Face and Gesture Recognition*, 1995, 1996, 1998, 2000, 2002.

[2]   AVBPA: *Proceedings of International Conference on Audio, Video and Biometric Person Authentication*, 1997, 1999, 2001, 2003.

[3]   J. J. Atick, P. A. Griffin, and A. N. Redlich, Statistical approach to shape from shading: Reconstruction of 3-dimensional face surfaces from single 2-dimensional images. *NeurComp*, **8**(6):1321–1340, Aug. 1996.

[4]   D. Ayers and R. Chellappa, Scenario recognition from video using a hierarchy of dynamic belief networks. In *Proc. of Intl. Conf. Pattern Recognition*, 2000, pp. 835–838.

[5]   A. Azarbayejani, T. Starner, B. Horowitz, and A. P. Pentland, Visually controlled graphics. *IEEE Trans. Pattern Anal. Machine Intell.*, **15**:602–605, June 1993. doi:10.1109/34.216730

[6]   C. D. Barclay, J. E. Cutting, and L. T. Kozlowski, Temporal and spatial factors in gait perception that influence gender recognition. *Perception Psychophy.*, **23**:145–152, 1978.

[7]   P. Barral, G. Dorme, and D. Plemenos, Visual understanding of a scene by automatic movement of a camera. In *Int. Conf. GraphiCon'99*, 1999.

[8]   R. Basri and D. W. Jacobs, Lambertian reflectance and linear subspaces. *IEEE Trans. Pattern Anal. Machine Intell.*, **25**(2):218–233, Feb. 2003. doi:10.1109/TPAMI.2003.1177153

[9]   T. Beardsworth and T. Buckner, The ability to recognize oneself from a video recording of ones movements without seeing ones body. *Bull. Psychonomic Soc.*, **18**(1):19–22, 1981.

[10] P. N. Belhumeur, J. P. Hespanha, and D. J. Kriegman, Eigenfaces vs. Fisher-faces: Recognition using class specific linear projection. *IEEE Trans. Pattern Anal. Machine Intell.*, **19**:711–720, 1997.doi:10.1109/34.598228

[11] I. Biederman and P. Kalocsai, Neural and psychophysical analysis of object and face recognition. In *Face Recognition: From Theory to Applications*, H. Wechsler, P. J. Phillips, V. Bruce, F. F. Soulie, and T. S. Huang, Eds. New York: Springer-Verlag, 1998, pp. 3–25.

[12] A. Bissacco, A. Chiuso, Y. Ma, and S. Soatto, Recognition of human gaits. In *Proc. of IEEE Computer Society Conf. Computer Vis. Pattern Recogn.*, Vol. 2, 2001, pp. 52–57.

[13] M. J. Black and Y. Yacoob, Tracking and recognizing rigid and non-rigid facial motions using local parametric models of image motion. In *Proc. Int. Conf. Comput. Vis.*, 1995, pp. 374–381.

[14] D. Blackburn, M. Bone, and P. J. Phillips, Face recognition vendor test 2000. Technical report available at http://www.frvt.org, 2001.

[15] V. Blanz and T. Vetter, Face recognition based on fitting a 3d morphable model. *IEEE Trans. Pattern Anal. Machine Intell.*, **25**(9):1063–1074, Sept. 2003. doi:10.1109/TPAMI.2003.1227983

[16] W. W. Bledsoe, The model method in facial recognition. Technical Report PRI:15, Panoramic Research, Inc., Palo Alto, CA, 1964.

[17] A. F. Bobick and A. Johnson, Gait recognition using static activity-specific parameters. In *Proc. IEEE Conf. Comput. Vis. Pattern Recogn.*, Dec. 2001, pp. 423–430.

[18] E. Borovikov and L. Davis, 3D shape estimation based on density driven model fitting. In *Proc. 1st Int. Symp. 3D Data Process. Visual. Trans. (3DPVT)*, 2002.

[19] M. Brand, Morphable 3D models from video. In *Proc. IEEE Comput. Soc. Conf. Comput. Vis. Pattern Recogn.*, 2001, pp. 456–463.

[20] M. Brand and R. Bhotika, Flexible flow for 3D nonrigid tracking and shape recovery. In *Proc. IEEE Comput. Soc. Conf. Comput. Vis. Pattern Recogn.*, 2001, pp. 315–322.

[21] C. Bregler, A. Hertzmann, and H. Biermann, Recovering non-rigid 3D shape from image streams. In *Proc. IEEE Comput. Soc. Conf. Comput. Vis. Pattern Recogn.*, 2000, pp. 690–696.

[22] C. Bregler and J. Malik, Tracking people with twists and exponential maps. In *Proc. IEEE Comput. Soc. Conf. Comput. Vis. Pattern Recogn.*, 1998, pp. 8–15.

[23] B. F. Bremond and M. Thonnat, Analysis of human activities described by image sequences. In *Proc. Intl. Florida AI Research Symp.*, 1997.

[24] A. Burton-Jones, V.C. Storey, V. Sugumaran, and P. Ahluwalia, Assessing the effectiveness of the DAML ontologies for the semantic web. In *Eighth Int. Conf. Appl. Natural Language to Inform. Syst.*, June 2003.

[25] H. Buxton and S. Gong, Visual surveillance in a dynamic and uncertain world. *Artif. Intell.*, 1995, Vol. **78**: 431–459.

[26] R. Cutler, C. Benabdelkader and L.S. Davis, Motion based recognition of people in eigengait space. In *Proc. IEEE Conf. Face Gesture Recogn.*, 2002, pp. 267–272.

[27] C. Castel, L. Chaudron, and C. Tessier, What is going on? A high-level interpretation of a sequence of images. In *ECCV Workshop on Conceptual Descriptions from Images*, 1996.

[28] T. Choudhury, B. Clarkson, T. Jebara, and A. Pentland, Multimodal person recognition using unconstrained audio and video. In *Proc. Intl. Conf. Audio, Video Biometric Person Authentication*, 1999, pp. 176–181.

[29] J. J. Clark and A. L. Yuille, *Data Fusion for Sensory Information Processing Systems*. Norwell, MA: Kluwer, 1990.

[30] K. D. Cock and D. B. Moor, Subspace angles and distances between ARMA models. In *Proc. Intl. Symp. Math. Theory Networks Syst.*, 2000.

[31] R. Collins, R. Gross, and J. Shi, Silhouette-based human identification from body shape and gait. In *Proc. IEEE Int. Conf. Automatic Face Gesture Recogn.*, May 2002, pp. 351–356.

[32] T. F. Cootes, C. J. Taylor, D. H. Cooper, and J. Graham, Active shape models: Their training and application. *Comput. Vis. Image Understand.*, **61**:38–59, Jan. 1995. doi:10.1006/cviu.1995.1004

[33]   I. J. Cox, J. Ghosh, and P. N. Yianilos, Feature-based face recognition using mixture-distance. In *Proc. IEEE Comput. Soc. Conf. Comput. Vis. Pattern Recogn.*, 1996, pp. 209–216.

[34]   D. Cunado, J. M. Nash, M. S. Nixon, and J. N. Carter, Gait extraction and description by evidence-gathering. In *Proc. Int. Conf. Audio, Video Biometric Person Authentication*, 1999, pp. 43–48.

[35]   J. Cutting and L. Kozlowski, Recognizing friends by their walk: Gait perception without familiarity cues. *Bull. Psychonomic Soc.*, **9**:353–356, 1977.

[36]   J. E. Cutting and D. R. Proffitt, Gait perception as an example of how we perceive events. In *Intersensory Perception and Sensory Integration*, R. D. Walk and H. L. Pick, Ed., London: Plenum Press, 1981.

[37]   J. Davis and A. Bobick, The representation and recognition of action using temporal templates. In *Proc. IEEE Comput. Soci. Conf. Comput. Vis. Pattern Recogn.*, 1997, pp. 928–934.

[38]   A. Doucet, N. deFreitas, and N. Gordon, *Sequential Monte Carlo Methods in Practice*. New York: Springer, 2001.

[39]   C. Dousson, P. Gabarit, and M. Ghallab, Situation recognition: Representation and algorithms. In *Proc. Int. Joint Conf. AI*, 1993, pp. 166–172.

[40]   I. L. Dryden and K. V. Mardia, *Statistical Shape Analysis*. New York: John Wiley, 1998.

[41]   R. Duda, P. Hart, and D. Stork, *Pattern Classification*, 2nd ed. New York: John Wiley, 2001.

[42]   G. J. Edwards, C. J. Taylor, and T. F. Cootes, Learning to identify and track faces in image sequences. In *Proc. Int. Conf. Comput. Vis.*, 1998, pp. 317–322.

[43]   H. D. Ellis, Introduction to aspects of face processing: Ten questions in need of answers. In *Aspects of Face Processing*, H. Ellis, M. Jeeves, F. Newcombe, and A. Young, Eds. Dordrecht: Nijhoff, 1986, pp. 3–13.

[44]   K. Etemad and R. Chellappa, Discriminant analysis for recognition of human face images. *J. Opt. Soc. Amer. A*, **14**:1724–1733, 1997.

[45]   O. D. Faugeras, *Three-Dimensional Computer Vision: A Geometric Viewpoint.* Cambridge, MA: MIT Press, 1993.

[46]   D. Forsyth, Shape from texture and integrability. In *Proc. Int. Conf. Comput. Vis.*, 2001, pp. 447–453.

[47]   R. T. Frankot and R. Chellappa, A method for enforcing integrability in shape from shading problem. *IEEE Trans. Pattern Anal. Machine Intell.*, **10**(7):439–451, 1987.

[48]   W. T. Freeman and J. B. Tenenbaum, Learning bilinear models for two-factor problems in vision. In *Proc. IEEE Comput. Soc. Conf. Comput. Vis. Pattern Recogn.*, 1997.

[49]   P. Fua, Regularized bundle-adjustment to model heads from image sequences without calibration data. *Int. J. Comput. Vis.*, **38**(2):153–171, July 2000. doi:10.1023/A:1008105802790

[50]   K. Fukunaga, *Statistical Pattern Recognition.* New York: Academic Press, 1989.

[51]   S. Furui, Cepstral analysis technique for automatic speaker verification. *Proc. Intl. Conf. Acous., Speech Signal Process.*, **29**(2):254–272, April 1981. doi:10.1109/TASSP.1981.1163530

[52]   I. Gauthier and N. K. Logothetis. Is face recognition so unique after all? *J. Cognit. Neuropsychol.*, **17**:125–142, 2000. doi:10.1080/026432900380535

[53]   D. M. Gavrila, The visual analysis of human movement: A survey. *Comput. Vis. Image Understand.*, **73**(1):82–98, Jan. 1999. doi:10.1006/cviu.1998.0716

[54]   S. Geman and D. Geman, Stochastic relaxation, gibbs distributions, and the Bayesian restoration of images. *IEEE Trans. Pattern Anal. Machine Intell.*, **6**(6):721–741, Nov. 1984.

[55]   A. Georghiades, P. Belhumeur, and D. Kriegman, From few to many: Illumination cone models for face recognition under variable lighting and pose. *IEEE Trans. PAMI*, **23**(6):643 –660, 2001.

[56]   G. Golub and C. V. Loan, *Matrix Computations.* Baltimore, MD: Johns Hopkins University Press, 1989.

[57] S. Gong, S. McKenna, and A. Psarrou, *Dynamic Vision: From Images to Face Recognition.* Singapore: World Scientific, 2000.

[58] W. E. L. Grimson, L. Lee, R. Romano, and C. Stauffer, Using adaptive tracking to classify and monitor activities in a site. In *Proc. IEEE Comput. Soc. Conf. Comput. Vis. Pattern Recogn.*, 1998, pp. 22–31.

[59] R. Gross and Jianbo Shi, The CMU motion of body (MOBO) database. Technical Report CMU-RI-TR-01-18, Robotics Institute, Carnegie Mellon University, Pittsburgh, PA, June 2001.

[60] T. R. Gruber, Toward principles for the design of ontologies used for knowledge sharing. *Int. J. Human-Comput. Stud.*, 1995.

[61] G. Shakhnarovich, L. Lee, and T. Darrell, Integrated face and gait recognition from multiple views. In *Proc. IEEE Comput. Soc. Conf. Comput. Vis. Pattern Recogn.*, Dec. 2001, pp. 439–446.

[62] L. Gu, S. Z. Li, and H. J. Zhang, Learning probabilistic distribution model for multi-view face dectection. In *Proc. IEEE Comput. Soc. Conf. Comput. Vis. Pattern Recogn.*, 2001.

[63] I. Haritaoglu, D. Harwood, and L. S. Davis, Ghost: A human body part labeling system using silhouettes. In *Proc. Int. Conf. Pattern Recogn.*, 1998, pp. 77–82.

[64] G. F. Harris and P. A. Smith, Ed. *Human Motion Analysis: Current Applications and Future Directions.* Piscataway, NJ: IEEE Press, 1996.

[65] R. I. Hartley and A. Zisserman, *Multiple View Geometry in Computer Vision.* Cambridge, UK: Cambridge University Press, 2000.

[66] B. Heisele, T. Serre, M. Pontil, and T. Poggio, Component-based face detection. In *Proc. IEEE Comput. Soc. Conf. Comput. Vis. Pattern Recogn.*, 2001, pp. 657–662.

[67] L. Hong and A. Jain, Multimodal biometrics. In *Biometrics: Personal Identification in a Networked Society*, A. Jain, R. Bolle, and S. Pankanti, Ed., Norwell, MA: Kluwer, 1999.

[68] S. Hongeng and R. Nevatia, Multi-agent event recognition. In *Proc. Int. Conf. Comput. Vis.*, 2001, pp. 84–91.

[69] B. K. P. Horn and M. J. Brooks, *Shape from Shading*. Cambridge, MA: MIT Press, 1989.

[70] P. S. Huang, C. J. Harris, and M. S. Nixon, Recognizing humans by gait via parametric canonical space. *Artif. Intell. Eng.*, **13**(4):359–366, Oct. 1999. doi:10.1016/S0954-1810(99)00008-4

[71] T. Huang, D. Koller, J. Malik, G. Ogasawara, B. Rao, S. Russell, and J. Weber, Automatic symbolic traffic scene analysis using belief networks. In *Proc. AAAI*, 1994, pp. 966–972.

[72] S. Intille and A. Bobick, A framework for recognizing multi-agent action from visual evidence. In *Proc. AAAI*, 1999, pp. 518–525.

[73] Michael Isard and Andrew Blake, Contour tracking by stochastic propagation of conditional density. *Proc. Eur. Conf. Comput. Vis.*, **1**: 343–356, 1996.

[74] T. Jebara, K. Russell, and A. Pentland, Mixtures of eigen features for real-time structure from texture. In *Proc. Int. Conf. Comput. Vis.*, 1998, pp. 128–135.

[75] T. S. Jebara and A. P. Pentland, Parameterized structure from motion for 3D adaptive feedback tracking of faces. In *Proc. IEEE Comput. Soc. Conf. Comput. Vis. Pattern Recogn.*, 1997, pp. 144–150.

[76] G. Johansson, Visual perception of biological motion and a model for its analysis. *PandP*, **14**(2):201–211, 1973.

[77] G. Johansson, Visual motion perception. *Sci. Amer.*, **232**:76–88, 1975.

[78] I. Kakadiaris and D. Metaxas, Model-based estimation of 3D human motion. *IEEE Trans. Pattern Anal. Machine Intell.*, 2000, **22**(12):1453–1459. doi:10.1109/34.895978

[79] A. Kale, A. Roy-Chowdhury, and R. Chellappa, Towards a view invariant gait recognition algorithm. In *Proc. IEEE Conf. Advanced Video Signal Based Surveillance*, 2003, pp. 143–150.

[80] A. Kale, N. Cuntoor, B. Yegnanarayana, A. N. Rajagopalan, and R. Chellappa, Gait analysis for human identification. In *Proc. Int. Conf. Audio, Video Biometric Person Authentication*, 2003.

[81] A. Kale, A. K. Roy-Chowdhury, and R. Chellappa, Gait-based human identification from a monocular video sequence. In *Handbook on Pattern Recognition and Computer Vision*, 3rd eds., C. H. Cheng and P. S. P. Wang, Ed. Singapore: World Scientific, to be Published.

[82] T. Kanade, *Computer Recognition of Human Faces*. Basel: Birkhauser, 1973.

[83] M. D. Kelly, Visual identification of people by computer. Technical Report AI-130, Stanford AI Project, Stanford, CA, 1970.

[84] D. G. Kendall, D. Barden, T. K. Carne, and H. Le, *Shape and Shape Theory*. New York: John Wiley, 1999.

[85] M. Kirby and L. Sirovich, Application of the Karhunen–Loeve procedure for the characterization of human faces. *IEEE Trans. Pattern Anal. Machine Intell.*, **12**:103–108, Jan. 1990. doi:10.1109/34.41390

[86] J. Kittler, M. Hatef, R. P. W. Duin, and J. Matas, On combining classifiers. *IEEE Trans. Pattern Anal. Machine Intell.*, **20**(3):226–239, March 1998. doi:10.1109/34.667881

[87] B. Knight and A. Johnston, The role of movement in face recognition. *Vis. Cogn.*, **4**:265–274, 1997.

[88] L. Kozlowski and J. Cutting, Recognizing the sex of a walker from a dynamic point display. *Perception Psychophy.*, **21**:575–580, 1977.

[89] N. Kruger, M. Potzsch, and C. von der Malsburg, Determination of face position and pose with a learned representation based on labeled graphs. *Image Vis. Comput.*, **15**:665–673, Aug. 1997. doi:10.1016/S0262-8856(97)00012-7

[90] Y. Kuniyoshi and H. Inoue, Qualitative recognition of ongoing human action sequences. In *Proc. Int. Joint Conf. AI*, 1993, pp. 1600–1609.

[91] M. Lades, J. Vorbruggen, J. Buhmann, L. Lange, C. von der Malsburg, R. Wurtz, and W. Konen. Distortion invariant object recognition in

the dynamic link architecture. *IEEE Trans. Comput.*, **42**:300–311, 1993. doi:10.1109/12.210173

[92] A. Lanitis, C. J. Taylor, and T. F. Cootes, Automatic face identification system using flexible appearance models. *Image Vis. Comput.*, **13**:393–401, June 1995. doi:10.1016/0262-8856(95)99726-H

[93] A. Laurentini, The visual hull concept for silhouette-based image understanding. *IEEE Trans. Pattern Anal. Machine Intell.*, **16**(2):150–162, Feb. 1994. doi:10.1109/34.273735

[94] K. C. Lee, J. Ho, M. H. Yang, and D. J. Kriegman, Video-based face recognition using probabilistic appearance manifolds. In *Proc. IEEE Comput. Soc. Conf. Comput. Vis. Pattern Recogn.*, 2003, pp. 313–320.

[95] L. Lee and G. Dalley, Learning pedestrian models for silhouette refinement. In *Proc. Int. Conf. Comput. Vis.*, 2003, pp. 663–670.

[96] L. Lee and W. E. L. Grimson, Gait analysis for recognition and classification. In *Proc. IEEE Int. Conf. Automatic Face Gesture Recogn.*, 2002, pp. 155–161.

[97] M. Levoy and P. Hanrahan, Light field rendering. In *Proc. ACM SIGGRAPH*, New Orleans, LA, 1996.

[98] B. Li and R. Chellappa, Face verification through tracking facial features. *J. Opt. Soc. Amer. A*, **18**(12):2969–2981, Dec. 2001.

[99] Y. M. Li, S. G. Gong, and H. Liddell, Constructing facial identity surfaces in a nonlinear discriminating space. In *Proc. IEEE Comput. Soc. Conf. Comput. Vis. Pattern Recogn.*, 2001, pp. 258–263.

[100] Y. M. Li, S. G. Gong, and H. Liddell, Modelling faces dynamically across views and over time. In *Proc. Int. Conf. Comput. Vis.*, 2001, pp. 554–559.

[101] S. H. Lin, S. Y. Kung, and L. J. Lin, Face recognition/detection by probabilistic decision-based neural network. *IEEE Trans. Neural Networks*, **8**:114–132, 1997. doi:10.1109/72.554196

[102] J. Little and J. Boyd, Recognizing people by their gait: The shape of motion. *Videre*, **1**(2):1–32, 1998.

[103]   C. Liu and H. Wechsler, Evolutionary pursuit and its application to face recognition. *IEEE Trans. Pattern Anal. Machine Intell.*, **22**:570–582, June 2000. doi:10.1109/34.862196

[104]   X. Liu and T. Chen, Video-based face recognition using adaptive hidden Markov models. In *Proc. IEEE Comput. Soc. Conf. Comput. Vis. Pattern Recogn.*, 2003, pp. 340–345.

[105]   J. Matas *et. al*, Comparison of face verification results on the XM2VTS database. In *Proc. Int. Conf. Pattern Recogn.*, Vol. 4, 2000, pp. 858–863.

[106]   W. Matusik, C. Buehler, and L. McMillan, Polyhedral visual hulls for real-time rendering. In *Proc. Eurographics Workshop Rendering'01*, 2001.

[107]   W. Matusik, C. Buehler, R. Raskar, S. J. Gortler, and L. McMillan, Image-based visual hulls. In *Proc. SIGGRAPH 2000*, pp. 369–374.

[108]   T. Maurer and C. von der Malsburg, Single-view based recognition of faces rotated in depth. In *Proc. IEEE Int. Conf. Automatic Face Gesture Recogn.*, 1996, pp 176–181.

[109]   S. McKenna and S. Gong, Recognising moving faces. In *Face Recognition: From Theory to Applications*, H. Wechsler, P. J. Phillips, V. Bruce, F. F. Soulie, and T. S. Huang, Eds. Berlin: Springer-Verlag, 1998, pp. 578–588.

[110]   L. McMillan, An image-based approach to 3D computer graphics. Ph.D. Dissertation, University of North Carolina, 1997.

[111]   K. Messer, J. Matas, J. Kittler, J. Luettin, and G. Maitre, Xm2vtsdb: The extended M2VTS database. In *Proc. Int. Conf. Audio, Video Biometric Person Authentication*, 1999, pp. 72–77.

[112]   B. Moghaddam, T. Jebara, and Pentland A, Bayesian face recognition. *Pattern Recogn.*, **33**:1771–1780, 2000. doi:10.1016/S0031-3203(99)00179-X

[113]   B. Moghaddam and A. P. Pentland, Probabilistic visual learning for object representation. *IEEE Trans. Pattern Anal. Machine Intell.*, **19**:696–710, July 1997. doi:10.1109/34.598227

[114] H. Moon, R. Chellappa, and A. Rosenfeld, Optimal edge based shape detection. *IEEE Trans. Image Process.*, **11**(11):1209–1227, Nov. 2002. doi:10.1109/TIP.2002.800896

[115] M. P. Murray, A. B. Drought, and R. C. Kory, Walking patterns of normal men. *J. Bone Joint Surg.*, **46-A**(2):335–360, 1964.

[116] E. Muybridge, *The Human Figure in Motion*. New York: Dover, 1901.

[117] H. Nagel, From image sequences towards conceptual descriptions. *Image Vis. Comput.*, **6**:59–74, 1988. doi:10.1016/0262-8856(88)90001-7

[118] V. Nalwa, *A Guided Tour of Computer Vision*. Reading, MA: Addison-Wesley, 1993.

[119] B. Neumann and H.J. Novak, Event models for recognition and natural language descriptions of events in real-world image sequences. In *Proc. Int. Joint Conf. AI*, 1983, pp. 724–726.

[120] R. Nevatia, J. Hobbs, and R. Bolles, An ontology for representing video events. In *IEEE Workshop Event Mining—Detection Recogn. of Events in Video*, 2004.

[121] P.V. Overschee and B.D. Moor, Subspace algorithms for the stochastic identification problem. *Automatica*, **29**:649–660, 1993. doi:10.1016/0005-1098(93)90061-W

[122] A. Papoulis, *Probabbility, Random Variables and Stochastic Processes*. New York: McGraw-Hill, 1991.

[123] V. Parameswaran and R. Chellappa, Quasi-invariants for human action representation and recognition. In *Proc. Int. Conf. Pattern Recogn.*, 2002, pp. 307–310.

[124] V. Parmeswaran and R. Chellappa, View invariants for human action recognition. In *Proc. IEEE Comput. Soc. Conf. Comput. Vis. Pattern Recogn.*, 2003, pp. 613–619.

[125] J. Pearl, *Probabilistic Reasoning in Intelligent Systems*. San Mateo, CA: Morgan Kaufmann, 1988.

[126] P. Penev and J. Atick, Local feature analysis: A general statistical theory for object representation. *Network: Comput. Neural Syst.*, **7**:477–500, 1996. doi:10.1088/0954-898X/7/3/002

[127] A.P. Pentland, B. Moghaddam, and T. Starner, View-based and modular eigenspaces for face recognition. In *Proc. IEEE Comput. Soc. Conf. Comput. Vis. Pattern Recogn.*, 1994.

[128] P. J. Phillips, S. Sarkar, I. Robledo, P. Grother, and K. W. Bowyer, Baseline results for the challenge problem of human ID using gait analysis. In *Proc. IEEE Int. Conf. Automatic Face Gesture Recogn.*, 2002, pp. 130–135.

[129] P. J. Phillips, S. Sarkar, I. Robledo, P. Grother, and K. W. Bowyer, The gait identification challenge problem: Data sets and baseline algorithm. In *Proc Int. Conf. Pattern Recogn.*, 2002, pp. 385–388.

[130] P.J. Phillips, P.J. Grother, R.J. Micheals, D.M. Blackburn, E. Tabassi, and J.M. Bone, Face recognition vendor test 2002: Evaluation report. Technical Report NISTIR 6965, http://www.frvt.org, 2003.

[131] P.J. Phillips, H. Moon, S. Rizvi, and P. Rauss, The FERET evaluation methodology for face-recognition algorithms. *IEEE Trans. Pattern Anal. Machine Intell.*, **22**:109–1104, Oct. 2000. doi:10.1109/34.879790

[132] P.J. Phillips, H. Wechsler, J. Huang, and P.J. Rauss, The FERET database and evaluation procedure for face-recognition algorithms. *Image Vis. Comput.*, **16**:295–306, April 1998. doi:10.1016/S0262-8856(97)00070-X

[133] M. Pollefeys, Self-calibration and metric 3D reconstruction from uncalibrated image sequences. Ph.D. Thesis, ESAT-PSI, K.U. Leuven, 1999.

[134] L.R. Rabiner, A tutorial on hidden Markov models and selected applications in speech recognition. *Proc. IEEE*, **77**(2):257–285, Feb. 1989. doi:10.1109/5.18626

[135] C. Rao, A. Yilmaz, and M. Shah, View-invariant representation and recognition of actions. *Int. J. Comput. Vis.*, **50**(2):203–226, 2002. doi:10.1023/A:1020350100748

[136] P. Remagnini, T. Tan, and K. Baker, Agent-oriented annotation in model based visual surveillance. In *Proc. of International Conf. on Computer Vision*, pages 857–862, 1998.

[137] H. Robbins and S. Monro, A stochastic approximation method. *Ann. Math. Stat.*, **22**:400–407, 1951.

[138] S. Romdhani, V. Blanz, and T. Vetter, Face identification by fitting a 3D morphable model using linear shape and texture error functions. In *Proc. Eur. Conf. Comput. Vis.*, IV: 3ff, 2002.

[139] N. Rota and M. Thonnat, Activity recognition from video sequence using declarative models. In *Eur. Conf. Artif. Intell.*, 2000.

[140] H.A. Rowley, S. Baluja, and T. Kanade, Neural network based face detection. *IEEE Trans. Patt. Anal. Machine Intell.*, **20**:23–38, Jan. 1998. doi:10.1109/34.655647

[141] A. K. Roy-Chowdhury, Statistical analysis of 3D modeling from monocular video streams. Ph.D. Thesis, Univeristy of Maryland, College Park, 2002.

[142] A. K. Roy-Chowdhury and R. Chellappa, A factorization approach for event recognition. In *CVPR Event Mining Workshop*, 2003.

[143] A. K. Roy-Chowdhury and R. Chellappa, Stochastic approximation and rate-distortion analysis for robust structure and motion estimation. *Int. J. Comput. Vis.*, **55**(1):27–53, Oct. 2003. doi:10.1023/A:1024488407740

[144] A.K.Roy-ChowdhuryandR.Chellappa,Aninformationtheoreticcriterion for evaluating the quality of 3D reconstructions from video. *IEEE Trans. Image Process.*, **13**(7):960–973, July 2004. doi:10.1109/TIP.2004.827240

[145] A. K. Roy-Chowdhury and R. Chellappa, Face reconstruction from monocular video using uncertainty analysis and a generic model. *Comput. Vis. Image Understand.*, **91**:188–213, July 2003. doi:10.1016/S1077-3142(03)00079-1

[146] H. Schneiderman and T. Kanade, Probabilistic modelling of local appearance and spatial reationships for object recognition. In *Proc. IEEE Comput. Soc. Conf. Comput. Vis. Pattern Recogn.*, 2000, pp. 746–751.

[147] W. I. Scholhorn, B.M. Nigg, D. J. Stephanshyn, and W. Liu, Identification of individual walking patterns using time discrete and time continuous data sets. *Gait Posture*, **15**:180–186, 2002. doi:10.1016/S0966-6362(01)00193-X

[148] G. Shaffer, *A Mathematical Theory of Evidence*. Princeton, NJ: Princeton University Press, 1976.

[149] Y. Shan, Z. Liu, and Z. Zhang, Model-based bundle adjustment with application to face modeling. In *Proc. Int. Conf. Comput. Vis.*, 2001, pp. 644–651.

[150] A. Shashua, Geometry and photometry in 3D visual recognition. Ph.D. Thesis, MIT, 1994.

[151] A Shashua, On photometric issues in 3D visual recognition from a single 2d image. *Int. J. Comput. Vis.*, **21**:99–122, 1997. doi:10.1023/A:1007975506780

[152] I. Shimshoni, Y. Moses, and M. Lindenbaum, Shape reconstruction of 3D bilaterally symmetric surfaces. *Int. J. Comput. Vis.*, **39**:97–100, 2000. doi:10.1023/A:1008118909580

[153] H. Sidenbladh, Michael J. Black, and David J. Fleet, Stochastic tracking of 3D human figures using 2D image motion. In *Proc. Eur. Conf. Comput. Vis.*, 2000, pp. 702–718.

[154] T. Sim, S. Baker, and M. Bsat, The CMU pose, illumination, and expression (PIE) database. In *Proc. IEEE Int. Conf. Automatic Face Gesture Recogn.*, 2002, pp. 46–51.

[155] S. Soatto, G. Doretto, and Y.N. Wu, Dynamic textures. In *Proc. Int. Conf. Comput. Vis.*, Vol. 2, 2001, pp. 439–446.

[156] S. Soatto and A.J. Yezzi, Deformotion: Deforming motion, shape average and the joint registration and segmentation of images. In *Proc. Euro. Conf. Comput. Vis.*, III: 32 ff, 2002.

[157] S. Srinivasan, Extracting structure from optical flow using fast error search technique. *Int. J. Comput. Vis.*, **37**:203–230, 2000. doi:10.1023/A:1008111923880

[158] T. Starner, J. Weaver, and A. Pentland, Real-time American sign language recognition from video using hmms. *IEEE Trans. Pattern Anal. Machine Intell.*, **12**(8):1371–1375, Dec. 1998. doi:10.1109/34.735811

[159] J. Steffens, E. Elagin, and H. Neven, Personspotter—Fast and robust system for human detection, tracking and recognition. In *Proc. IEEE Int. Conf. Automatic Face Gesture Recogn.*, 1998, pp. 516–521.

[160] S. V. Stevenage, M. S. Nixon, and K. Vince, Visual analysis of gait as a cue to identity. *Appl. Cogn. Psychol.*, **13**:513–526, March 1999. doi:10.1002/(SICI)1099-0720(199912)13:6<513::AID-ACP616>3.0.CO;2-8

[161] J. Strom, T. Jebara, S. Basu, and A.P. Pentland, Real time tracking and modeling of faces: An EKF-based analysis by synthesis approach. In *Vismod*, 1999.

[162] A. Sundaresan, A. K. Roy-Chowdhury, and R. Chellappa, A hidden Markov model based framework for recognition of humans from gait sequences. In *Proc. Int. Conf. Image Process.*, Sept. 2003.

[163] A. Sundaresan, A. K. Roy-Chowdhury, and R. Chellappa, Multi-view tracking of human motion modeled by kinematic chains. In *Proc. Int. Conf. Image Process.*, 2004.

[164] K. Sung and T. Poggio, Example-based learning for view-based human face detection. *IEEE Trans. Pattern Anal. Machine Intell.*, **20**:39–51, Jan. 1997. doi:10.1109/34.655648

[165] D.L. Swets and J. Weng, Discriminant analysis and eigenspace partition tree for face and object recognition from views. In *Proc. IEEE Int. Conf. Automatic Face Gesture Recogn.*, 1996, pp. 192–197.

[166] D.L. Swets and J.J. Weng, Using discriminant eigenfeatures for image retrieval. *IEEE Trans. Pattern Anal. Machine Intell.*, **18**:831–836, Aug. 1996. doi:10.1109/34.531802

[167] R. Tanawongsuwan and A.F. Bobick, Gait recognition from time-normalized joint-angle trajectories in the walking plane. In *Proc. IEEE Comput. Soc. Conf. Comput. Vis. Pattern Recogn.*, 2001, pp. 726–731.

[168] D. Terzopoulos and K. Waters, Analysis and synthesis of facial image sequences using physical and anatomical models. *IEEE Trans. Pattern Anal. Machine Intell.*, **15**:569–579, June 1993. doi:10.1109/34.216726

[169] D. Tolliver and R. Collins, Gait shape estimation for identification. In *Proc. Int. Conf. Audio, Video Biometric Person Authentication*, 2003, pp. 734–742.

[170] C. Tomasi and T. Kanade, Shape and motion from image streams under orthography: A factorization method. *Int. J. Comput. Vis.*, **9**:137–154, Nov. 1992. doi:10.1007/BF00129684

[171] L. Torresani and C. Bregler, Space-time tracking. In *Proc. Eur. Conf. Comput. Vis.*, I: 801 ff, 2002.

[172] S. Tsuji, A. Morizono, and S. Kuroda, Understanding a simple cartoon film by a computer vision system. In *Proc. Int. Joint Conf. AI*, 1977, pp. 609–610.

[173] M. Turk and A. Pentland, Eigenfaces for recognition. *J. Cogn. Neurosci.*, **3**:72–86, 1991.

[174] V.N. Vapnik, *The Nature of Statistical Learning Theory*. New York: Springer-Verlag, 1995.

[175] N. Vaswani, A. R. Chowdhury, and R. Chellappa, Statistical shape theory for activity modeling. In *Proc. Int. Conf. Acoustics, Speech Signal Process.*, 2003.

[176] N. Vaswani, A. R. Chowdhury, and R. Chellappa, Activity recognition using the dynamics of the configuration of interacting objects. In *Proc. IEEE Comput. Soc. Conf. Comput. Vis. Pattern Recogn.*, 2003, pp. 633–640.

[177] N. Vaswani, A. R. Chowdhury, and R. Chellappa, Shape activity: A continuous state hmm for moving/deforming shapes with application to abnormal activity detection. *IEEE Trans. Image Process.*, to be published.

[178] A. Veeraraghavan, A. R. Chowdhury, and R. Chellappa, Role of shape and kinematics in human movement analysis. In *Proc. IEEE Comput. Soc. Conf. Comput. Vis. Pattern Recogn.*, 2004.

[179] V.T. Vu, F. Bremond, and M. Thonnat, Automatic video interpretation: A recognition algorithm for temporal scenarios based on pre-compiled scenario models. In *CVS03*, 2003, p. 523.

[180] R. Walter, *Principles of Mathematical Analysis*, 3rd ed. New York: McGraw-Hill, 1976.

[181] H. Wechsler, V. Kakkad, J. Huang, S. Gutta, and V. Chen, Automatic video-based person authentication using the RBF network. In *Proc. Int. Conf. Audio, Video Biometric Person Authentication*, 1997, pp. 85–92.

[182] J. Wilder, Face recognition using transform coding of gray scale projection and the neural tree network. In *Artificial Neural Networks with Applications in Speech and Vision*, R.J. Mammone, Ed. New York: Chapman Hall, 1994, pp. 520–536.

[183] A. Wilson and A Bobick, Recognition and interpretation of parametric gesture. In *Proc. Int. Conf. Comput. Vis.*, 1998, pp. 329–336.

[184] L. Wiskott, J. M Fellous, and C. von der Malsburg, Face recognition by elastic bunch graph matching. *IEEE Trans. Pattern Anal. Machine Intell.*, **19**:775–779, 1997. doi:10.1109/34.598235

[185] M. Yamamoto and K. Koshikawa, Human motion analysis based on a robot arm model. In *Proc. IEEE Comput. Soc. Conf. Comput. Vis. Pattern Recogn.*, 1991, pp. 664–665.

[186] M. Yamamoto, A. Sato, S. Kawada, T. Kondo, and Y. Osaki, Incremental tracking of human actions from multiple views. In *Proc. IEEE Comput. Soc. Conf. Comput. Vis. Pattern Recogn.*, 1998, pp. 2–7.

[187] G.S. Young and R. Chellappa, Statistical analysis of inherent ambiguities in recovering 3D motion from a noisy flow field. *IEEE Trans. Pattern Anal. Machine Intell.*, **14**:995–1013, Oct. 1992. doi:10.1109/34.159903

[188] A. L. Yuille, D. Snow, Epstein R., and P. N. Belhumeur, Determining generative models of objects under varying illumination: Shape and albedo from multiple images using svd and integrability. *Int. J. Comput. Vis.*, **35**:203–222, 1999. doi:10.1023/A:1008180726317

[189] A.L. Yuille, D.S. Cohen, and P.W. Hallinan, Feature extraction from faces using deformable templates. *Int. J. Comput. Vis.*, **8**:99–111, Aug. 1992. doi:10.1007/BF00127169

[190] L. Zhao, Dressed human modeling, detection, and parts localization. Ph.D. Thesis, CMU, 2001.

[191] W. Zhao, Robust image based 3D face recognition. Ph.D. Thesis, University of Maryland, 1999.

[192] W. Zhao and R. Chellappa, Symmetric shape from shading using self-ratio image. *Int. J. Comput. Vis.*, **45**:55–752, 2001. doi:10.1023/A:1012369907247

[193] W. Zhao, R. Chellappa, and A. Krishnaswamy, Discriminant analysis of principal components for face recognition. In *Proc. IEEE Int. Conf. Automatic Face Gesture Recogn.*, 1998, pp. 336–341.

[194] W. Zhao, R. Chellappa, P.J. Phillips, and A. Rosenfeld, Face recognition: A literature survey. *ACM Trans.*, **35**:399–458, Dec. 2003.

[195] S. Zhou and R. Chellappa, Illuminating light field: Image-based face recognition across illuminations and poses. In *Proc. IEEE Int. Conf. Automatic Face Gesture Recogn.*, May 2004.

[196] S. Zhou and R. Chellappa, Image-based face recognition under illumination and pose variation. *J. Opt. Soc. Amer. A*, **22**:217–229, Feb. 2005, submitted for publication.

[197] S. Zhou and R. Chellappa, Multiple-exemplar discriminant analysis for face recognition. In *Proc. of Intl. Conf. Pattern Recogn.*, Aug. 2004.

[198] S. Zhou, R. Chellappa, and D. Jacobs, Characterization of human faces under illumination variations using rank, integrability, and symmetry constraints. In *Eur. Conf. Comput. Vis.*, May 2004, pp. 588–601.

[199] S. Zhou, V. Krueger, and R. Chellappa, Probabilistic recognition of human faces from video. *Comput. Vis. Image Understand.*, **91**:214–245, July–Aug. 2003. doi:10.1016/S1077-3142(03)00080-8

The Authors

RAMA CHELLAPPA

Rama Chellappa received the B.E. (Hons.) degree from the University of Madras, India, in 1975 and the M.E. (Distinction) degree from the Indian Institute of Science, Bangalore, in 1977. He received the M.S.E.E. and Ph.D. degrees in electrical engineering from Purdue University, West Lafayette, IN, in 1978 and 1981, respectively.

Since 1991 he has been a Professor of Electrical Engineering and an Affiliate Professor of Computer Science at the University of Maryland, College Park. He is also affiliated with the Center for Automation Research (Director) and the Institute for Advanced Computer Studies (permanent member). Prior to joining the University of Maryland, he was an Assistant (1981–1986) and Associate Professor (1986–1991) and Director of the Signal and Image Processing Institute (1988–1990) with the University of Southern California, Los Angeles. Over the last 23 years he has published numerous book chapters, peer reviewed journal and conference papers. He has edited a collection of papers on Digital Image Processing (published by IEEE Computer Society Press), coauthored (with Y.T. Zhou) a research monograph on Artificial Neural Networks for Computer Vision, published by Springer-Verlag, and coedited (with A.K. Jain) a book on Markov Random fields, published by Academic Press. His current research interests are face and gait analysis, 3D modeling from video, automatic target recognition from stationary and moving platforms, surveillance and monitoring, hyper spectral processing, image understanding, and commercial applications of image processing and understanding.

Dr. Chellappa has served as an associate editor of the *IEEE Transactions on Signal Processing, Pattern Analysis and Machine intelligence, Image Processing, and*

Neural Networks. He was a coeditor-in-chief of *Graphical Models and Image Processing.* He also served as the editor-in-chief of *IEEE Transactions on Pattern Analysis and Machine Intelligence* during 2001–2004. He has also served as a member of the IEEE Signal Processing Society Board of Governors during 1996–1999 and as its Vice President of Awards and Membership during 2002–2004. He has received several awards, including NSF Presidential Young Investigator Award, an IBM Faculty Development Award, the 1990 Excellence in Teaching Award from the School of Engineering at USC, the 1992 Best Industry Related Paper Award from the International Association of Pattern Recognition (with Q. Zheng), and the 2000 Technical Achievement Award from the IEEE Signal Processing Society. He was elected as a Distinguished Faculty Research Fellow (1996–1998) and as a Distinguished Scholar–Teacher (2003) at the University of Maryland. He is a fellow of the International Association for Pattern Recognition. He has served as a General the Technical Program Chair for Server IEEE international and national conferences and workshops.

AMIT K. ROY-CHOWDHURY

Amit K. Roy-Chowdhury received the B.E. degree in electrical engineering from Jadavpur University, India, in 1995; the M.E. degree in systems science and automation from the Indian Institute of Science, Bangalore, in 1997; and Ph.D. from the University of Maryland, College Park, in 2002. His Ph.D. thesis was on statistical error characterization of 3D modeling from monocular video sequences. His current research interests are in motion and illumination modeling in video sequences, computational models for human activity recognition, and video sensor networks.

Since 2004 he has been an Assistant Professor in the Department of Electrical Engineering, University of California, Riverside. In 2003 he was with the Center for Automation Research, University of Maryland, College Park, as a Research Associate, where he worked in projects related to face, gait, and activity recognition. He is the author of a number of papers, book chapters, and magazine articles on motion analysis, 3D modeling, and object recognition. He is a reviewer for journals in signal and image processing and computer vision and has served on the Program Committees of major international conferences in these areas. He was a receipient of University of California Regents' Faculty Fellowship Award in 2004–2005.

S. KEVIN ZHOU

S. Kevin Zhou is a Research Scientist at Siemens Corporate Research, Princeton, New Jersey. He received his B.E. degree in electronic engineering from the University of Science and Technology of China, Hefei, China, in 1994; M.E. degree in computer engineering from the National University of Singapore in 2000; and Ph.D. in electrical engineering from the University of Maryland at College Park in 2004. He has broad research interests in signal/image/video processing, computer vision, pattern recognition, machine learning, and statistical inference and computing. He has published more than 30 technical papers and book chapters on echocardiography image processing, visual recognition (in particular face recognition under unconstrained conditions, such as video sequences, illumination, and pose variations, etc.), tracking and motion analysis, segmentation and shape/appearance modeling, learning under uncertainty, and optimization and efficient computation.

Printed in the United States
by Baker & Taylor Publisher Services